READ MY PINS

MADELEINE ALBRIGHT

READ MY PINS

STORIES FROM A DIPLOMAT'S JEWEL BOX

MADELEINE ALBRIGHT

with
Elaine Shocas, Vivienne Becker, and Bill Woodward

Photography by John Bigelow Taylor
Photography Composition by Dianne Dubler

HARPER

An Imprint of HarperCollinsPublishers
www.harpercollins.com

See a pin, pick it up,
And all day you'll have good luck.
See a pin, let it lay,
And your luck will pass away.

— Nursery Rhyme

With deep appreciation to
St. John Knits for its support
of the book and to Bren Simon
for her support of the exhibition.

This book is published in conjunction
with the exhibition "Read My Pins:
The Madeleine Albright Collection"
organized by the Museum of Arts and
Design, New York. After being shown
at the Museum, the exhibition will tour
to selected venues in the United States
and around the world.

HarperCollins books may be purchased for educational, business, or sales promotional use. For information, please write: Special Markets Department, HarperCollins Publishers, 10 East 53rd Street, New York, NY 10022.

This book was produced by Melcher Media, 124 West 13th Street, New York, NY 10011, www.melcher.com.
PUBLISHER: Charles Melcher
ASSOCIATE PUBLISHER: Bonnie Eldon
EDITOR IN CHIEF: Duncan Bock
PROJECT EDITORS: Lindsey Stanberry and David E. Brown
PRODUCTION DIRECTOR: Kurt Andrews
PRODUCTION ASSISTANT: Daniel del Valle

Pin photography copyright © 2009 John Bigelow Taylor. Further photography credits are listed on page 173.

Designed by Rita Jules, Miko McGinty, Inc.

Every attempt has been made to identify and contact the designer or manufacturer of each pin and to list its country of origin or, if unknown, the country in which it was acquired. The date listed in the pindex reflects the year or estimated year the pin was made or, if unknown, the year it was acquired. The centimeter dimensions have been converted to the nearest tenth of an inch. Any errors brought to the publisher's attention will be corrected in future editions.

ISBN: 978-0-06-089918-9

FIRST EDITION

Library of Congress Cataloging-in-Publication data is available upon request.

09 10 11 12 13   10 9 8 7 6 5 4

PIN CAPTIONS:
Page 1: The Great Seal of the United States book locket and pin, Ann Hand.
Page 2: The United States Capitol, Monet.
Page 4, clockwise from top left: Asymmetrical gold heart, Erwin Pearl; red heart and bow, Ann Hand; bejeweled heart, designer unknown; sparkling red heart, Ann Hand; interlocking hearts, Swarovski; purple heart, D.M. Lee; hammered metal heart, Omega; rhinestone bombé heart, designer unknown.
Page 7, clockwise from top left: Gold ginkgo leaf, designer unknown; silver ginkgo leaf, designer unknown; copper ginkgo leaf, Dennis Ray/Beauvoir, the National Cathedral Elementary School; gold-stemmed ginkgo leaf, Fabrice.
Page 9: Victory Knot, Verdura.
Page 11: Alert Lady, Brit Svenni/Berit Kowalski. According to the designers, "One eye is extra watchful as Madeleine Albright is always alert to the world's problems."
Pages 12–13, row 1, from left: Black rhinestone butterfly, Ann Hand; green and coral butterfly, Kenneth Jay Lane; blue butterfly, designer unknown; light blue rhinestone butterfly, Ciner; blue enamel butterfly, designer unknown; row 2: large silver butterfly, Christian Dior; gold butterfly, Cécile et Jeanne; lattice filigree butterfly, Caviar; opal butterfly, Tiny Jewel Box; pearl butterfly, Kenneth Jay Lane; row 3: gold butterfly and wreath, Miriam Haskell; amber butterfly, designer unknown; green and violet butterfly, Modital Bijoux; rhinestone butterfly, José & María Barrera; silver and blue butterfly, designer unknown; gray rhinestone butterfly, Ciner.

Table of Contents

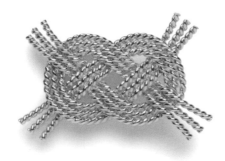

INTRODUCTION

In *Six Memos for the Next Millennium*, the great Italian short story–writer Italo Calvino recounts the legend of how the emperor Charlemagne was enchanted by a gold ring. Whoever or whatever possessed the ring held the power of bewitchment, from a deceased maiden to an archbishop and, ultimately, the lake into which the ring was cast. This small piece of jewelry took on magical powers, becoming "an outward and visible sign that reveals the connection between people or between events." In addition to conveying information about the wearer—her or his status, finances, and affinities—jewelry has an impressive power to establish links among people, places, and events. Madeleine Albright's pins are nothing if not eloquent and often provocative communicators.

Secretary Albright's pins cannot be described as a collection in any traditional sense. Collectors usually set out with specific goals in mind as to what they intend to acquire and how they will secure the objects of desire, whether they be paintings, stamps, butterflies, or grandfather clocks. By contrast, Secretary Albright's collection has grown organically over the years in response to the changing circumstances and opportunities of her life and career. This is a collection that has been amplified and enriched by the events that have engaged their owner, providing a visible record of past experiences and future hopes.

There is a delightful randomness and whimsy to the pins that make up this highly personal assemblage. Sought out in settings ranging from jewelry stores and art galleries to airport souvenir stands and the booths of craft fair vendors, they first spoke to Secretary Albright, asking (sometimes demanding) to be included in her trove of wearable images. Their value as communication devices once recognized, they were then inducted into service as diplomatic aides; sometimes demure and understated, sometimes outlandish and outspoken, they became gentle implements of statecraft.

The pins reveal a rich diversity of motifs and images. Angels, stars, balloons, American flags, and spaceships are juxtaposed with a menagerie of birds, bees, beetles, butterflies, fish, frogs, turtles, and snakes. A variety of garden flowers, sentimental hearts and bows, and mementos of specific events and holidays round out the collection.

Jewelry buffs typically focus their attention on the preciousness of the materials from which an item is made—gold, silver, rubies, or diamonds—or on the virtuosity of the craftsmanship revealed in its design. Secretary Albright's pins, however, are for the most part unremarkable in their monetary value and, except for some pieces of antique or fine jewelry, likely to be by anonymous designers, and fabricated from materials ranging from base metals to plastics and glass. Rhinestones and crystal take the lead roles over diamonds, electroplating over solid gold.

Of modest intent and manufacture, Secretary Albright's pins are of a kind that anyone could possess and wear. These are truly "pins of the people," and part of Secretary Albright's pleasure in wearing the pins must come from her recognition of their democratic nature. To assemble so notable a collection of pins takes something much more elusive and significant than money—it takes a magical combination of a collector's eye, which can spot and home in on its target, and an ability to recognize the communicative potential of what might be deemed ordinary things. Through her pins, Secretary Albright tells us a great deal about herself—her sense of humor and her humanity—and does so with grace and flair.

It is especially gratifying to know that this delightful collection, with its engaging history and purpose, can be shared with so many through this publication and the memorable exhibition it accompanies.

David Revere McFadden
Chief Curator, Museum of Arts and Design, New York

I. The Serpent's Tale

The idea of using pins as a diplomatic tool is not found in any State Department manual or in any text chronicling American foreign policy. The truth is that it would never have happened if not for Saddam Hussein.

During President Bill Clinton's first term (1993–1997), I served as America's ambassador to the United Nations. This was the period following the first Persian Gulf War, when a U.S.-led coalition rolled back Iraq's invasion of neighboring Kuwait. As part of the settlement, Iraq was required to accept UN inspections and to provide full disclosure about its nuclear, chemical, and biological weapons programs.

Above: Voting in the UN Security Council. That is the serpent pin on my jacket.

Page 14: The pin that began it all. *Serpent, designer unknown.*

When Saddam Hussein refused to comply, I had the temerity to criticize him. The government-controlled Iraqi press responded by publishing a poem entitled "To Madeleine Albright, Without Greetings." The author, in the opening verse, establishes the mood: "Albright, Albright, all right, all right, you are the worst in this night." He then conjures up an arresting visual image: "Albright, no one can block the road to Jerusalem with a frigate, a ghost, or an elephant." Now thoroughly warmed up, the poet refers to me as an "unmatched clamor-maker" and an "unparalleled serpent."

In October 1994, soon after the poem was published, I was scheduled to meet with Iraqi officials. What to wear?

Years earlier, I had purchased a pin in the image of a serpent. I'm not sure why, because I loathe snakes. I shudder when I see one slithering through the grass on my farm in Virginia. Still, when I came across the serpent pin in a favorite shop in Washington, D.C., I couldn't resist. It's a small piece, showing the reptile coiled around a branch, a tiny diamond hanging from its mouth.

While preparing to meet the Iraqis, I remembered the pin and decided to wear it. I didn't consider the gesture a big deal and doubted that the Iraqis even made the connection. However, upon leaving the meeting, I encountered a member of the UN press corps who was familiar with the poem; she asked why I had chosen to wear that particular pin. As the television cameras zoomed in on the brooch, I smiled and said that it was just my way of sending a message.

A second pin, this of a blue bird, reinforced my approach. As with the snake pin, I had purchased it because of its intrinsic appeal, without any extraordinary use in mind. Until the twenty-fourth of February 1996, I wore the pin with the bird's head soaring upward. On the afternoon of that tragic day, Cuban fighter pilots shot down two unarmed civilian aircraft over international waters between

Right: I used blunt words to express anger and sadness when, in 1996, airplanes carrying four Cuban-American fliers were shot down off the coast of Florida. My blue bird pin reflected my mood.

Opposite page: Blue bird, Anton Lachmann.

Cuba and Florida. Three American citizens and one legal resident were killed. The Cubans knew they were attacking civilian planes yet gave no warning, and in the official transcripts they boasted about destroying the *cojones* of their victims.

At a press conference, I denounced both the crime and the perpetrators. I was especially angered by the macho celebration at the time of the killings. "This is not *cojones*," I said, "it is cowardice." To illustrate my feelings, I wore the bird pin with its head pointing down, in mourning for the free-spirited Cuban-American fliers. Because my comment departed from the niceties of normal diplomatic discourse, it caused an uproar in New York and Washington; for the same reason, it was welcomed in Miami. As a rule, I prefer polite talk, but there are moments when only plain speaking will do.

To Our Secretary
On a happy
day in Texas
with warm
best wishes
Gg Bush

Left: My first trip as secretary of state was to Texas, where I was greeted warmly by former President George H.W. Bush and Barbara Bush at their Houston home. In the background is Millie, their pet springer spaniel, author of *Millie's Book: As Dictated to Barbara Bush,* a 1991 best-seller. My pin, hard to see, is an eagle with a pearl.

Below: Sun, Steinmetz Diamonds.

Before long, and without intending it, I found that jewelry had become part of my personal diplomatic arsenal. Former President George H.W. Bush had been known for saying, "Read my lips." I began urging colleagues and reporters to "Read my pins."

While teaching at Georgetown University, I tell my students that the purpose of foreign policy is to persuade others to do what we want or, better yet, to want what we want. To accomplish this, a president or secretary of state has a range of tools that includes military force at one end of the spectrum and words of reason at the other. In between are such instruments as diplomacy, economic sanctions, foreign aid, and trade. Compared to these, the brooch or pin may seem trivial.

I do not claim too much, but I do believe the right symbol at the correct time can add warmth or needed edge to a relationship. A foreign dignitary standing alongside me at a press conference would be happier to see a bright, shining sun attached to my

Opposite page: Crystal fly, Christian Dior; green and blue rhinestone bees, Ciner; turquoise bee, Walter Lampl; golden bee, St. John Knits.

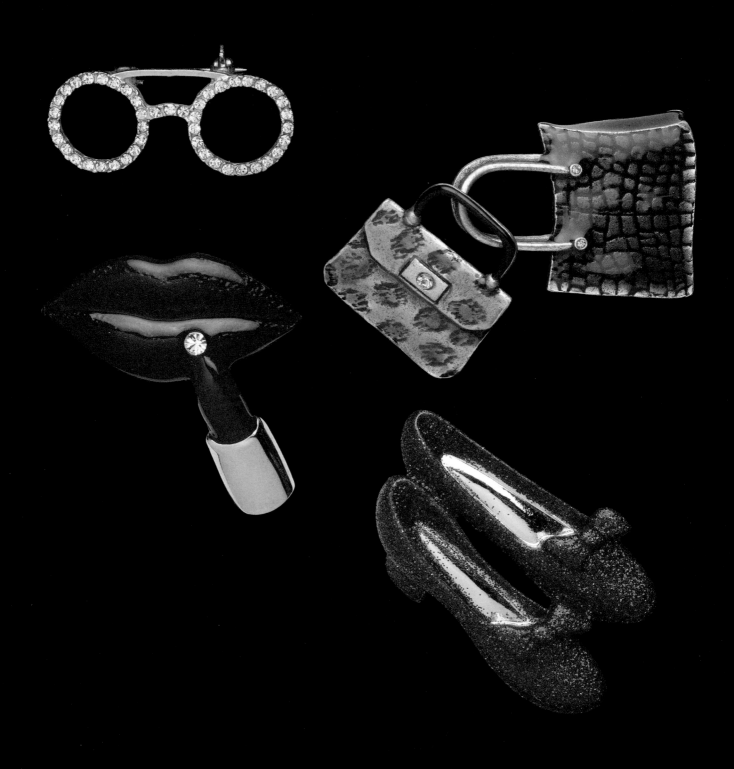

Above: Petit Oiseau, Jacqueline Lecarme.

Opposite page: Pins need not cost a king's or queen's ransom to be fun. These gifts from a friend were less than three dollars each. Leopard and reptile print purses, AJMC; Ruby Slippers, AJC; other designers unknown.

jacket than a menacing wasp. I felt it worthwhile, moreover, to inject an element of humor and spice into the diplomatic routine. The world has had its share of power ties; the time seemed right for the mute eloquence of pins with attitude.

Because many of my predecessors had beards and none was known to wear a skirt, my use of pins to send a message was something new in American diplomacy. The role of jewelry in world affairs, however, began in ancient times. Throughout history, jewelry has played a supporting role in the rise and fall of empires. Although I might display to the world my less-than-extravagant pins, the global audience has long gaped in amazement at the stunning ornamentation of royal necks, waists, wrists, arms, and ankles— and at the accompanying crowns, thrones, scepters, and swords. To the victor go the spoils, and often those spoils have glistened with the radiance of diamonds or the soft glow of emeralds.

Laurel wreath, designer unknown.

Although monarchs typically tried to hoard their treasure, the demands of politics often prompted them to make use of it. Early diplomatic practices included the exchange of ornamental gifts between one head of state and another, the gift of jewelry to cement marriages that brought two nations into alliance, and the flaunting of riches for the purpose of engendering awe.

Consider, for example, the story of one of the first international power couples: Marc Antony and Queen Cleopatra. According to the near-contemporary account of Pliny the Elder, Cleopatra wagered with Antony that she could spend an extravagant amount of wealth on a single dinner. He accepted the bet. The following

Antony and Cleopatra at table. The pearl is about to drop. *The Banquet of Marc Antony and Cleopatra*, by Francesco Trevisani (1656–1746).

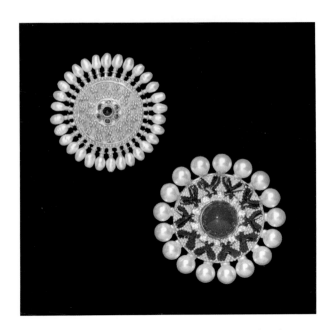

Left: Byzantine shield, Ilias Lalaounis; right: circle of pearls, Craft.

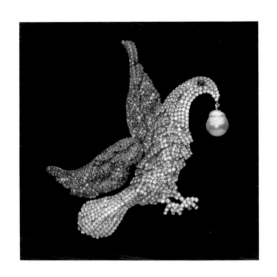

Bird with pearl, Bettina von Walhof.

night, she served a meal of conventional dishes, to which he responded with triumphant disdain. Smiling at the vanity of her Roman suitor, Cleopatra ordered the next course to be brought in: a single cup of strong vinegar. Deftly removing one of her priceless pearl earrings, the queen dropped it into the vinegar, causing the gem to dissolve.

A millennium later and several thousand miles to the north, a less elegant effort to connect the fates of two kingdoms was attempted. Olaf Tryggvason, the warrior king of Norway, set out to woo Sigrid, the comely queen of Sweden. Both had had previous romantic entanglements. Olaf had murdered a local rival named Iron Beard, claiming his victim's daughter for a wife. The daughter, displeased with the arrangement, spoiled the honeymoon by stabbing Olaf in bed. Divorce ensued. For her part, Sigrid had grown weary of two boorish suitors. One night, she allowed the duo to drink themselves into unconsciousness before locking the beer hall and torching the building. Thereafter, the queen was known as Sigrid the Strong-minded.

The betrothal of Olaf and Sigrid made sense diplomatically at a time and in a region where those without allies rarely prospered. Thus the queen was willing and the king eager enough to send his

prospective bride a beautiful gold ring. Sigrid's fancy, though, was of the type bred more in head than in heart. She promptly sent the ring to her goldsmiths for appraisal. The experts had only to lift the object to know there was something rotten in Norway; sure enough, the gold band on the outside concealed copper on the inside. In diplomacy, as in love, cheapskates rarely prosper. Olaf's ring was rejected, and Sigrid married the king of Denmark.

From early times, the rulers of India were encouraged to accumulate stockpiles of jewels to enhance their reputations and to outpace potential rivals. They did so not only through taxation but also through wars of conquest, supported by plunder. It is unsurprising, therefore, that the word "loot" is derived from a Hindi verb (*lūt*). When, in 1293, a Venetian traveler visited the king of Malabar in southern India, he found a man so wealthy that even his loincloth (set with emeralds, sapphires, and rubies) was worth a fortune. During the Mogul Empire, the men in a maharaja's court wore ornate necklaces, bracelets, and rings; the court's horses and elephants were outfitted with golden tassels and helmets, jewel-laden saddles and anklets, and—in the case of elephants—gold bands around their tusks.

Such wealth did not go unnoticed by foreign visitors. The capitals of Renaissance Europe possessed dazzling assets but lacked one vital ingredient: an indigenous source of gems. The desire to establish a reliable supply line was a major contributor to the Age of Exploration. Christopher Columbus sailed west in search of the mysterious East, inspired by his heavily annotated copy of Marco Polo's journal, which promised mighty palaces "all roofed with the finest gold." Though he discovered no golden roofs, Columbus nevertheless thought he had reached Asia; sailors less eager for a shortcut actually did. Guiding their fleets around the Cape of Good Hope, these merchant adventurers established their presence along India's coast.

Above: Indian elephant, DeNicola.

Opposite page: Castle, LJ.

Above: Giving an important speech on Bosnia at the Intrepid Sea, Air, and Space Museum in New York, 1997. I wore a fleur-de-lis, then a part of Bosnia's state flag. As is evident from the Queen Mother's crown (*below*), the fleur-de-lis was also popular among European royalty. *Left: Rhinestone fleur-de-lis, designer unknown; gold fleur-de-lis, Sofie.*

For several centuries, European traders competed for the favor of the subcontinent's leading families. The difficulty was how to bribe rulers who were already so rich. For a time, Portuguese merchants had the advantage because their offerings were novel: Colombian emeralds and Mozambican gold, amber, and ivory. Frustrated suitors eventually realized, however, that gift-giving is not the only means of persuasion.

By the start of the nineteenth century, the power of the Mogul dynasty was waning just as Great Britain's might was waxing. The ambition of Queen Victoria, the increased strength of Her Majesty's navy, and the skill and aggression of English traders forced India into a role it never wanted: the jewel in the British Empire's crown. The decisive blow came when, in 1849, the East India Trading Company gained control of Lahore, capital of Punjab. Most prominent among the riches claimed by the company and forwarded as

State crown featuring the famous Koh-i-Noor diamond. *Courtesy of the Royal Collection, Her Majesty Queen Elizabeth II.*

The Hope diamond. *Courtesy of Smithsonian Institution National Museum of Natural History.*

a tribute to the Queen was an enormous diamond, the incomparable Koh-i-Noor, or "mountain of light."

In the British capital, crowds rushed to see the Koh-i-Noor, but many Londoners were disappointed at its seeming lack of brilliance. Similarly unimpressed, the Queen's consort, Prince Albert, ordered the piece recut. In little more than a month, the craftsmen produced a beautiful shallow oval. Victoria subsequently wore the polished diamond in a brooch, in a tiara, and as the center of a diadem fashioned by Garrard, the crown jeweler. The gem was later placed in a Maltese cross at the front of the crown of Queen Elizabeth—known to my generation as the Queen Mum—and displayed at her state funeral in the spring of 2002.

The British crown jewels exemplify the connection between perceptions of national glory and the appreciation of valuable stones. This linkage transcends the borders of time, religion, geography, and culture. A traveler today, with sufficient time and the right access, could view everything from the large Manchurian pearls of China's Qing dynasty to the treasures of the pharaohs, from the crown jewels of Ethiopia to the imperial possessions of the Holy Roman and Austro-Hungarian empires.

In my own travels, I have visited the Tower of London, the Hermitage in Saint Petersburg, the Museum of Egyptian Antiquities in Cairo, the National History Museum of Romania, and the Louvre, where what little remains of the French crown jewels is on display. The majority of the French pieces were either lost in the Revolution or sold later to discourage attempts at restoring the Bourbon dynasty. As one radical parliamentarian exclaimed, "Without a crown, no need for a king."

The United States, of course, has never desired a crowned head and thus has no crown jewels—though the Smithsonian Institution has the Hope diamond and other extraordinary gems. Early Americans provided proof that jewelry is not the province solely of

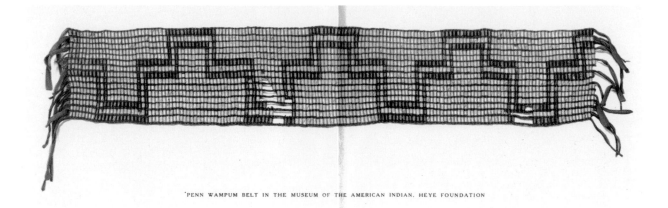

Above: This wampum belt, sometimes referred to as the "Freedom" belt, is presumed to have been given to William Penn by the Lenape, or Delaware, Nation, as early as 1682. *Courtesy of Smithsonian Institution National Museum of the American Indian.*

Above: A gift from two governors of Arizona, Janet Napolitano and Rose Mofford. *Western Sun, Federico Jimenez.*

Opposite page: Eagle Dancer, Jerry Roan.

royalty. American Indians were skilled at fashioning white, purple, and black beads out of the shells of periwinkles and clams. The beads, known as wampum, were used to record treaties and for other purposes both spiritual and practical. Like a royal crown, beaded headpieces, necklaces, and belts were employed by American tribes to connote leadership status; as with other jewels, wampum might be exchanged to acquire goods, express friendship, pay reparations, or facilitate peace.

For the New World's European settlers, wampum served as legal tender alongside the coins brought from their homelands. Ever alert for ways to push the natives aside, the settlers learned quickly that the more wampum they accumulated, the easier it would be to buy local land. In the most famous transaction, Peter Minuit, an employee of the Dutch West India Company, purchased Manhattan and later Staten Island for a modest amount of wampum, fabric, and farming implements. The Norwalk Indians accepted a comparable bargain in Connecticut, selling much of what is now Fairfield County.

As these examples suggest, jewelry has played a colorful part in the evolution of world affairs. Because precious stones tend to inspire both admiration and greed, leaders have found convenient excuses for seeking them and have used them to impress crowds, reward friends, deprive foes, forge alliances, and justify war. Jewels may find their highest expression in the decorative arts, but they have also earned a place in the art of the possible.

The role of jewelry in politics first touched my life at an early age. I was eight when my father served as ambassador from our native Czechoslovakia to Yugoslavia, then headed by Marshal Tito, a formidable dictator. During a diplomatic ceremony in Belgrade, my mother was invited to sit in an anteroom with the wives of two other ambassadors. Suddenly, the door opened and a Yugoslav fighter dressed in faded fatigues strode in bearing a silver tray. On the tray were three velvet boxes; in each was a ring made from the appropriate birthstone. The box presented to my mother—she was born in May—revealed an emerald surrounded by fourteen diamonds. We called it Tito's ring, and when my father first saw it, he growled, "I wonder whose finger they cut off to get this." Both my parents spoke of the contrast between the pomp and extravagance of the Yugoslav regime and the desperate poverty that plagued the country's people in those first years after World War II.

This was my mother's most valued pin, a gift from her sister.

Above: Tito's ring, designer unknown.

Left: Dignitaries gathered from around the world to attend the funeral of Yugoslav strongman Marshal Tito in Belgrade, 1980. I was standing off to the left, outside of the picture.

Above: My parents, Josef and
Mandula Korbel, during World War II.
Below: A pin of my mother's, designer
unknown.

Sometimes the finest jewelry is accompanied by moral complexity; there was no diplomatic way to return the gift. Instead, my parents waited until I had passed the orals for my Ph.D., then gave the ring to me.

In the late 1970s, I worked for Zbigniew Brzezinski, national security advisor to President Jimmy Carter. Part of my job was to report each morning on international developments that might warrant the president's attention. The death of a major foreign leader, such as Tito, fit that description. After months of reporting that Tito was ill; then gravely ill; possibly deceased; and then still alive, I was able to confirm that Tito was undeniably and reliably dead. Vice President Walter Mondale led the U.S. delegation to the funeral, and I—because of my childhood association with Yugoslavia—was invited to come along. After three decades, the moment was finally right to wear Tito's ring.

I was born in Prague, capital of Czechoslovakia, which later split into the Czech Republic and Slovakia. Both countries remain close to my heart. President Václav Havel, hero of the Velvet Revolution, is among the people I most admire. The art nouveau pins, opposite, are based on designs by Alphonse Mucha, a famed artist and Slavic nationalist of the early twentieth century. At right is the Order of the White Lion, an award I received, in 1997, from Havel and the Czech government.

II. Wings

In the fall of 1955, I enrolled at Wellesley, a women's college ensconced comfortably within one of the more distant and bucolic suburbs of Boston. The fifties were a period of transition for American women, and although the curriculum at Wellesley was modern, some of the customs were not. Many of my classmates arrived on campus as I did, decked out in the style of the day—with a camel-hair coat, Shetland sweater, Bermuda shorts, circle pin, and a single strand of pearls.

Wearing my mother's ring. High school photo, 1955.

Early on, we were sent to the physical education department to pose for what was called a posture picture. This was to see whether we had "an understanding of good body alignment and the ability to stand well." To ensure accuracy, we were not allowed to wear any clothing above the waist. If we flunked, we were made to do exercises. I always wondered what happened to the pictures, until a few years ago, when they were discovered in a vault . . . at Yale.

Wellesley women were, on the whole, excellent students, and many went on to have stellar careers. At the time, however, thoughts of history and philosophy competed with chemistry of a nonacademic sort. The majority of us hoped to be engaged before we graduated. According to the tradition, one became "pinned" while a junior and engaged as a senior before receiving—on the

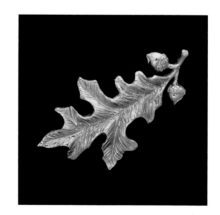

Alumnae leaf, Wellesley College.

Page 36: Bird, Iradj Moini.

afternoon of commencement day—a diploma at two o'clock and a wedding ring at four. Today, young women are more likely to get pierced than pinned, but back then we viewed the pinning ritual with great seriousness. A boy gave his fraternity pin to a girl, thereby pledging both affection and fidelity. When the girl wore the pin, on a blouse above her heart, she advertised that she was spoken for. The arrangement brought a couple's standing to a new and higher plane: more than dating, not always formally engaged.

The pin from Theta Delta Xi.

As a singularly mature and independent Wellesley woman, I was not the type to get married the same afternoon I graduated. Instead, I waited three days. I had met Joe Albright, my future husband, right on schedule in the summer between my sophomore and junior years; we both had jobs at the *Denver Post*. Joe, a fledgling reporter, was handsome in a tweedy way, but I kept my distance until verifying that the gold band on his finger was only a class ring. That issue resolved, we were immediately smitten, and within weeks Joe had proposed, offering his Theta Delta Xi pin to cement the deal. I was in heaven—but also in trouble, because I had another boyfriend, who knew nothing about Joe. Until I summoned the courage to break old ties, I kept Joe's pin out of sight, wearing it on my bra instead of my blouse. At first, only my sister, Kathy, knew of our plans for marriage. When Joe and I eventually told my parents, my father congratulated him for "pinning Madeleine down."

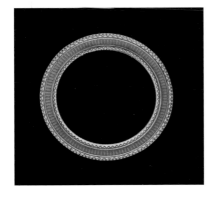

My circle pin.

Returning to Wellesley for my junior year, I virtually floated into the assembly hall at convocation wearing a red sweater accented by Joe's pin. My friends squealed appreciatively, and I promptly shared most, but not all, of the details of my summer romance and future plans. I quickly learned, though, that getting pinned and staying pinned were separate challenges. That winter, Joe had second thoughts about his career and decided that an early marriage might prove a hindrance. When he disclosed his

Above: Two arms full. My twins, Alice and Anne, 1961.

Below: Feather, designer unknown.

thinking, I was dumbfounded and began taking my pin off slowly, hoping Joe would stop me. He didn't, so I dropped the formerly precious object in his lap, whereupon he stood up, opened the window, and tossed the pin away. By the next morning, Joe had experienced third thoughts. Arriving at my room to accompany me to breakfast, he brought with him the keepsake he had tramped out into the snowy New England night to retrieve.

The following summer, Joe presented me with an antique emerald-and-diamond engagement ring he had bought in London. I loved it because it was beautiful but also because it was different. Other girls had engagement rings; I had *this* engagement ring. As I had discovered well after I had fallen in love with Joe, his family was socially prominent in both Chicago and New York. During our engagement, Joe's grandmother gave me a gorgeous antique jade pin; it was slender and long, decorated with a carved dragon. When Joe's sister Alice saw me wearing it, she looked as if she had swallowed a lemon. She told me later that she had first seen the piece while shopping with her grandmother and had pronounced it lovely. Naturally, she thought the dragon would one day come to her. I felt guilty—but not so much as to part with the pin.

Jade dragon, designer unknown.

Above: Ruby fish, designer unknown.

Right: Daughter Katie, soon after her christening, 1967.

Violets, designer unknown.

My wedding present from Joe was a fish pin with an emerald eye and ruby scales; on the back was the symbol for infinity. This hopeful hint of timelessness did not turn out to be apt, as our marriage ended in divorce after twenty-three years. In that time, I received an occasional gift but rarely shopped for jewelry myself. This was because women were expected to get their finery from men and because I was busy raising three children. I also never thought of the family money as mine to spend. The gifts, however, were much appreciated. From my grandmother-in-law (if there is such a thing), I received a second pin, this one feather-shaped, gold, with rubies. Joe's Uncle Harry Guggenheim gave my twin daughters little seed pearl hearts. Joe himself bought me a lapis lazuli turtle pin, a small brooch of a spray of violets, and a necklace of irregular pearls from Saudi Arabia that I wore all the time. There were also little gifts of trinkets and beads that were pretty but not built to last.

My most cherished jewelry: A heart pin made by Katie (*above*), and a gift from my parents, a Bohemian garnet set with detachable pendant/pin (*opposite page*).

The piece of jewelry that meant the most to me, then as now, was created by Katie, my youngest daughter. It is a heart-shaped pin, composed of clay, presented to me on Valentine's Day when Katie was five. I have often worn it since. The pin reflects one of the indispensable purposes of jewelry: to bind families together and connect one generation to the next. When I was a child, my sole treasures were a ring—a gold band with a single small diamond—that my mother had worn and a gold cross that I remember never being without. On my wedding day, my parents gave me a garnet set (a necklace, pin, bracelet, and earrings), featuring the Czechoslovak national stone. Usually, the cherished family gifts go from the elder to the younger, but as was the case with Katie's valentine, sometimes the giving goes the other way round.

After Joe's Aunt Alicia Patterson Guggenheim died, my daughters and I received a small share of her jewelry. This included a beautiful pink tourmaline heart and a diamond-and-sapphire poppy pin with matching earrings. There was also a pair of earrings with little pearls and a jade fish on the end that were meant to go with the jade dragon pin I had been given earlier. Although I adored these pieces, I so feared losing them that I rarely wore them. In any case, showy jewelry made me uncomfortable. Because of the social status of Joe's family, he had been considered the perfect escort for Chicago's well-bred young ladies, taking them to debutante balls and similar high society affairs. Suddenly he began to appear with me. Having nothing suitable to wear, I sewed a dark-red velvet dress to go with the garnets I had received from my parents. It had a tight waist and quite a low neckline so the garnets would show. I still have the dress as a reminder both of the evening and of the years when—for me—a tight waist was possible.

My life changed when Joe and I moved to Washington in the early 1960s. Jacqueline Kennedy, with her affinity for Givenchy and Oleg Cassini gowns and Schlumberger jewelry, was bringing unprecedented glamour to the nation's capital and America's global image. She wore diamonds to Paris, pearls to India, and bangles everywhere. Jackie, as she was called, was recognized as a fashion trendsetter, known by millions for her jewelry, handbags, hats, and hair. This was also the era of such spectacular movie icons as Elizabeth Taylor, Audrey Hepburn, and Marilyn Monroe, who, when singing "Diamonds Are a Girl's Best Friend," embodied the stereotype of a woman willing to be possessed but only in return for possessions.

Still, even in the swinging sixties, status in Washington was determined more by power than by glitter. My husband and I socialized with other young couples turned on by the promise and politics of the Kennedy administration. The men in our group

Poppy, Verdura.

mostly had jobs in government or as journalists; the women were active with children and social causes.

The jewelry worn by the wives in our circle consisted primarily of engagement and wedding rings, the occasional pearl necklace, and earrings that were generally nondescript but sometimes op art or pop art. We thought of jewelry as a traditional and fun means of adornment that was paid for by (usually male) acquaintances or that came to us through family ties. A fancier or more expensive item might make some statement about how much a husband could afford, but it was not a declaration of any depth about the woman wearing it.

Following my divorce in 1983, I found myself tapping into another sort of tradition. By then, I had completed my education and started out in politics. I had begun drawing a salary of my own working for a U.S. senator, Edmund S. Muskie of Maine, and then in the White House under Jimmy Carter, in whose honor I wore a pin shaped like a Georgia peanut. After that, I followed in

Above: Wrapped Heart, Verdura.

Opposite page: Meli Melo, Cartier.

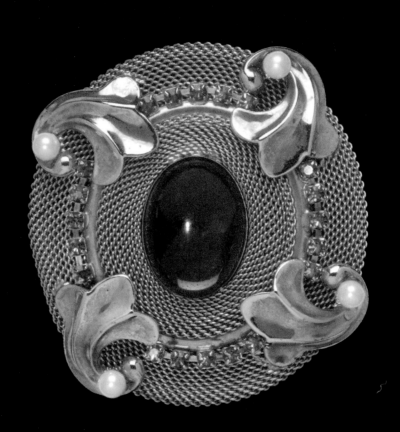

my father's footsteps, becoming a university professor. Though devastated when my marriage fell apart, I soon found my own spirit and voice. From that time on, when my mind turned to jewelry or clothes, I thought less about the expectations of others and more about my own sense of identity and pride.

My experiences, of course, were hardly unique. Women have been striding toward independence for many generations. In Great Britain in the early twentieth century, supporters of the suffragette movement wore medals or brooches in the shades of green, white, and violet—signifying, respectively, hope, purity, and dignity. Not coincidentally, the initial letters of the colors (*G, W, V*) suggested an acronym: "Give Women the Vote!" In the United States, suffragettes were equally flamboyant, greeting Woodrow Wilson's inauguration with an 8,000-person march down Pennsylvania Avenue led by a woman dressed as Joan of Arc and seated on a white horse. During Wilson's second term, activists who were thrown into jail for picketing were given a distinctive Jailed for Freedom brooch, produced by the indomitable women's-rights advocate Alice Paul. The pin displayed a prison door with a chain and heart-shaped padlock. In 1920, shortly before Wilson left office, the Nineteenth Amendment to the Constitution was ratified, and women were finally accorded the opportunity to vote in federal elections.

In my case, I didn't have a special color, didn't dress as Saint Joan, didn't go to jail, and didn't think of myself as belonging to a movement. Instead, I was in my mid-forties and venturing from marriage into—for the first time—the status of a fully grown, unattached adult. Although medals, ribbons, T-shirts, hats, and, I suppose, tattoos were optional means of expression, I found myself frequently turning to pins or brooches. I preferred them to necklaces because a perfectly presentable pin is less expensive than a comparable necklace. I also preferred pins because for years I did not want to wear a ring. In fact, the only one I felt comfortable

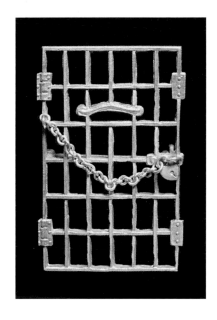

The silver Jailed for Freedom pin was awarded to suffragettes who were imprisoned after picketing in front of the White House in 1917. *Jailed for Freedom, Nina Evans Allender and Alice Paul. Courtesy of Smithsonian Institution National Museum of American History.*

Opposite page: Suffragette pin, designer unknown.

wearing was purchased in the Philippines and made of black onyx. My thinking at the time was that every divorced woman should wear a black ring.

The fashions of the 1980s have been described as postfeminist, which was fine with me since I had largely missed out on the earlier phases. The idea was that a woman could show independence from stereotypes without eschewing ornamentation; it was no longer thought essential to dress plainly in order to be taken seriously or to imply that wearing earrings made one unable to think. Since women were making inroads in business and the professions, power jackets and pantsuits came into style. The brooch was a natural accompaniment.

During my first decade of postmarital independence, I taught world affairs at Georgetown University and advised presidential candidates, most of whom lost. I also dated and shopped quite a bit. This is when I discovered the Tiny Jewel Box, a boutique situated on Connecticut Avenue in Washington's busiest commercial district. The store is actually more narrow than tiny. It advertises itself as "six intimate floors filled with treasures from around the world, each one hand-selected with an eye for the truly unique."

I have spent many an afternoon wandering about the Tiny Jewel Box's displays of antique pieces (typically bought from estates) and newer items by hot designers. My motives at the outset were entirely pure; I marched through the doors intent on selecting a necklace or brooch to give to a relative or friend in celebration of some event. If the occasion were a wedding, I might also decide to buy something elegant to wear at the ceremony; if a lesser event, a bauble to match a dress.

Before long, I accepted that it was okay to shop with my own needs and desires in mind. Thus, when my eye was attracted to a serpent pin, I did not hesitate to buy it; this was the pin that would later launch my use of brooches as a diplomatic tool.

Above: French urn, designer unknown.

Opposite page: The sheaf of wheat is a symbol of abundance and health. This pin was given to me upon my return to Georgetown University after my time as secretary of state. *Sheaf of wheat, Tiffany & Co.*

In a celebratory mood at Katie's wedding, joined by daughters
Alice and Anne, and my three sons-in-law, Greg Bowes
(holding grandson David), Jake Schatz, and Geoff Watson.
Grapes, Tiny Jewel Box.

Late in 1992, President-elect Clinton asked me to serve as America's ambassador to the United Nations. During the Cold War, the UN Security Council had been frozen by rivalry between East and West. The Council could only act when the superpowers agreed, and they did not agree often or about very much. When I arrived, three years after the collapse of the Berlin Wall, relations had thawed and the Council had new life. Instead of the big powers preventing cooperative action, they were asking the world body to take on jobs no country wanted to do alone. This had important implications for international law—and for my wardrobe.

Before leaving for New York, I consulted with my colleague at Georgetown, Jeane Kirkpatrick, who had been UN ambassador when Ronald Reagan was president. Kirkpatrick gave me one piece of advice: "Lose the professor clothes." Until then, I had been a student, mother, government staffer, and teacher; this was to be my first prolonged experience in the limelight. I spent time trying on outfits in various Washington boutiques; soon I also confirmed what no one has ever doubted: New York offers boundless opportunities to shop. As a friend from the Big Apple told me, "The only real difference between a human being and other mammals is our ability to accessorize."

My friend Jeane Kirkpatrick gave me good advice before my move to New York.

This pin was made from fragments of the Berlin Wall in 1989, the year the Wall was brought down. *Berlin Wall, Gisela Geiger.*

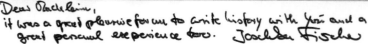

Surrounded by men at a meeting of the Group of Eight (G-8) Foreign Ministers. In the pin below, which represents the G-8, gender is not an issue.

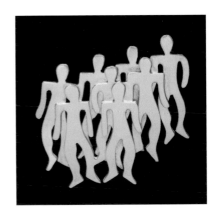

G-8 pin, designer unknown.

I had expected my initial Security Council meeting to be in the huge room with the horseshoe table that is frequently seen on television, but that chamber is generally reserved for formal sessions. The space used for routine meetings—where much of the real work is done—was no bigger than the college seminar rooms I had just left. The similarity reminded me of what I had frequently told my female students: Do not be afraid to interrupt. A woman usually prefers to size up a situation before speaking, but for America's UN ambassador, silence was not an option. So I squeezed my way into the cramped space, sat down, and, when the opportunity arose, plunged in.

From that day forward, I attracted attention because I represented the United States and was the only woman on the Council. With so many eyes on me, I didn't want to worry about my appearance. This prompted me to pay added heed to how I looked and gradually to acquire new pins to make my clothes more interesting. Because every activity at the United Nations has a political aspect, one of my signature themes was Americana.

New York's famed Pier Antiques Show is held periodically on the far West Side of midtown, in a sprawling building overlooking the Hudson River. Collectors and dealers from around the world are on hand, and whenever I could arrange my schedule, so was I. For seekers of high-quality costume jewelry, this was the equivalent of the Promised Land. Amid the crowd of shoppers, I moved from booth to booth, looking, touching, inquiring about prices, and—as one quickly learns to do in New York—using my elbows. One year, after surveying my options, I selected an eagle brooch manufactured by the distinguished American firm Trifari; it was enameled in red, white, and blue and set with rhinestones. Nearby, I came across an Uncle Sam's hat, also by Trifari, in a similar style. Both were made of enameled metal, and both dated from the 1940s. Either seemed suitable for America's UN ambassador, but I found the best effect came when I wore the two together, with the hat tilted at a rakish angle, seemingly atop the eagle's head (shown on page 59).

Above: Sailor, Monet.

Left: Overseas, accompanied by officers from the U.S. Air Force.

French president Jacques Chirac practicing the art of diplomacy. My pin (shown on the following page) celebrates *liberté, égalité, fraternité.*

Memorial bow, Trifari.

I also purchased a large American flag pin that I have since grown accustomed to wearing on the Fourth of July and other festive occasions. For funerals, to which I have been too often, I picked up a tricolor memorial bow.

The manufacture of costume jewelry with a patriotic theme flourished in the United States during and immediately after World War II. All the symbols I love—eagles, flags, drums, trumpets, and rousing slogans—were in vogue. The pieces were worn by noncombatants to signify support for the war effort and bought by sailors and soldiers to leave with sweethearts before taking up arms across the sea. Many of the pins came in the colors of the U.S. flag and continued selling after the war (except for those in red, which fell out of favor because of the color's association with Communism).

One of the reasons I appreciate costume jewelry is that it can delight the eye and still spare the pocketbook. The modern woman

I wore my Chinese shard dragon pin when testifying before Congress concerning U.S.-China relations.

Page 58, row 1, from left: American flag, Ann Hand; brave heart, Swarovski; AIDS ribbon, designer unknown; heart with donkey, designer unknown; heart stickpin, reproduction, The Metropolitan Museum of Art; row 2: French ribbon bow, Silson; patriotic bow, Carolee; Statue of Liberty, designer unknown; safety-pin American flag, designer unknown; row 3: small American flag, designer unknown; large Fight for Freedom torch, DMW; WWII ribbon, Silson; Stars and Stripes, Ciner; small Fight for Freedom torch, DMW. Page 59: Uncle Sam Top Hat and Eagle, Trifari.

Opposite page: One of my many bold pins. Colorful bird, Iradj Moini.

needs to be able to experiment with a look and try different ideas. Given my height (five foot two), I had always assumed small pins were best for me, but soon I began to buy pieces that—although not costly—were bigger, bolder, and sometimes even crazier. To my surprise, I found that the look I preferred was more on the dramatic side than the demure.

As my pins became more expressive and drew more comments, I had cause to reflect on the relationship between appearance and identity. To what extent, to adapt the old saying, do pins make the woman or, for that matter, the man? After all, the display of pins has never been confined to one gender. Medieval knights wore

Josiah Wedgwood's abolitionist medallion. *Courtesy of the Trustees of the British Museum.*

elaborate jeweled badges that defined their status and conferred a group identity. A fourteenth-century English lad could have no higher aspiration than to advertise a connection to the royal family by embellishing his cloak with the Order of the Garter's radiant star. Conspirators on all sides in the English Civil War used pins, rings, and lockets to signal their loyalties to friends without tipping off their enemies. George Washington sometimes wore a spectacular diamond eagle, based on a design by Pierre L'Enfant and given to him by the French Navy, that included no fewer than 198 precious stones. Pottery pioneer Josiah Wedgwood, Washington's contemporary, manufactured a medallion to be worn by opponents of the slave trade. Exquisitely carved, the cameo showed a black man in chains with the question, "Am I not a man and a brother?"

Above, from left: ESM pin, Cartier; saxophone, Kenneth Jay Lane; Solidarity, designer unknown.

Above, from left: Heart-health red dress, National Institutes of Health; Susan G. Komen for the Cure breast cancer ribbon.

Below: Harry Truman was president when my family arrived in America. Since then I have participated in many political campaigns and worn my share of buttons, including these featuring Truman and five other presidents.

In our own day, security experts rely on coded pins to identify people who are cleared to enter a particular area while excluding those who are not. Members of Congress are given pins so that they might avoid being stopped by guards while en route to their offices or the legislative floor. Clubs and lodges typically use badges (along with secret handshakes) to enable fellow members to recognize a common bond. Voters display pins to demonstrate allegiance to political candidates or causes. My own loyalties could be seen in a pin labeled "ESM," identifying me as an early backer of U.S. Senator Edmund S. Muskie; a "Solidarity" clip that, in 1981, I was given by anti-Communist activists in Poland; a pink ribbon for the race to cure breast cancer; a red dress for women's heart health; and a saxophone pin evoking our one and only sax-playing chief executive, Bill Clinton.

Finally, our armed forces also use pins—in the form of ribbons and medals—to convey messages about accomplishments, stature, and rank. I was reminded of this during my time as UN ambassador.

With General John Shalikashvili, who succeeded Colin Powell as chairman of the Joint Chiefs of Staff. In the background are Janet Langhart Cohen and Secretary of Defense Bill Cohen. It was Janet who designed the eagle and dove pin on the opposite page, celebrating NATO's fiftieth anniversary.

In 1993, the chairman of the Joint Chiefs of Staff was General Colin Powell. We were both members of the Clinton national security team and sat across from each other at the long rectangular table in the White House Situation Room. Although we saw eye to eye on many controversies, we did not agree on whether the United States and NATO should intervene to stop ethnic cleansing in the former Yugoslav Republic of Bosnia. I thought we should act and listed the reasons why; General Powell expressed doubts and listed the reasons why not. The dilemma for me was that although I had my patriotic pins, he had a chest full of well-earned medals. He was fresh from victory in the first Gulf War and cut a dashing figure in his uniform; I was fresh from my classrooms at Georgetown and, even in my best suit, resembled something of a dumpling (this was before I began working out).

As a civilian and a woman, I did not feel comfortable challenging the wisdom of such a true American hero, but I also knew that I had not been given a chair in the Situation Room to imitate a potted plant. I took my own advice and interrupted, arguing that the United States had an urgent interest in halting the slaughter of innocent people in the Balkans. Using a pointer and slides, Powell made clear his own expectation that the potential costs of such an effort would far outweigh the benefits. For months, we were at a stalemate; the administration did nothing, and, as the killing continued, I grew frustrated and at one meeting finally let loose.

Opposite page: Partners in Peace, Janet Langhart Cohen/Ann Hand.

Above: Some of my more deadly pins. *Jambiya dagger, Yemen; rocket-propelled grenade launcher (RPG), Pakistan.*

Left: On the cover of *Time* magazine, May 17, 1999.

"Colin," I asked, "what are you saving this superb military for, if we can't use it?" In his autobiography, Powell wrote that my question almost gave him an aneurysm and that he was compelled to explain to me—patiently—the appropriate role of the U.S. armed forces. In retrospect, I am willing to concede that the general was right to be cautious, right to ask questions, right to consider alternatives, and right to worry about the facile assumptions of civilian leaders. However, I was right about Bosnia, where NATO did eventually intervene and as a result saved thousands of lives.

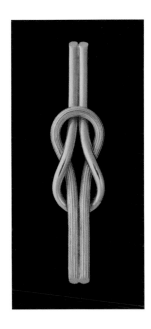

Knot of Hercules,
Ilias Lalaounis.

Although I love pins, they have in common with necklaces and bracelets one complication: the clasp. Thus it has ever been. The earliest pins were less ornamental than functional. Primitive hunter-gatherers used thorns or sharp pieces of flint to keep their clothes from falling off while they ran around in pursuit of lunch. As civilization progressed from the Stone Age to the Bronze and Iron eras, pieces of metal began to serve the same essential purpose. From there, it was only a short step to the use of rare ores and gemstones that combined the fastening with the alluring. Royal burial sites in Ur, home city of the patriarch Abraham, included gold and silver pins—some topped by lapis lazuli beads—that would have been used to secure robes at the shoulder.

The concept of the safety pin—in which a needle-like shaft, a hinge, and a sheath combine to secure an object—dates back to ancient Crete, the home turf of Theseus and his ill-tempered Minotaur. Metalsmiths in pre-Christian Etruria (present-day Tuscany) skillfully shaped such pins into the form of lions, horses, or the Sphinx before adding the frosting: tiny granules of gold. The brooch-clip, which clenches the fabric rather than piercing it, has

This gold brooch-clip is decorated with a winged chimaera, a beast that was popular in Etruscan mythology. Etruscan, circa 525–500 BC. *Courtesy of the Trustees of the British Museum.*

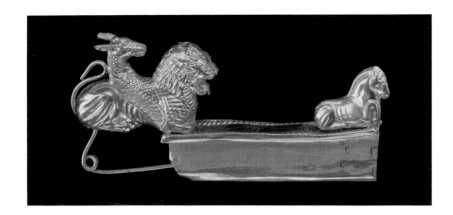

In December 1996, President Clinton nominated me to serve as America's sixty-fourth secretary of state. For the announcement, I wore one of my pins as a pendant. *Liberty Eagle, Ann Hand.*

been used widely since the 1930s. I cite this history to prove that I could not possibly have been the first person to be publicly embarrassed by a pin that came undone in a moment of need.

On January 23, 1997, shortly before noon, I was sworn in as secretary of state, the first woman to hold that position. Ever since, people have asked what I was feeling at the time. The answer is that my attention was divided between the drama of the moment and the possibility that my pin would fall off, landing on the floor in front of President Clinton and the assembled cameras. I had been introduced to the pin weeks earlier at the Tiny Jewel Box. Jim Rosen-heim, one of the proprietors, brought it to me as soon as I walked in, saying he had acquired the piece with me in mind. The brooch is antique, French, and composed of rose-cut diamonds and a gold eagle with widespread wings. It was love at first sight, but I balked at the cost. Saying no to Jim, I inwardly promised to reverse that decision should I be named secretary of state, then a possibility but hardly a likelihood.

When that possibility became reality, I bought the eagle and chose to wear it for the first time at the swearing in. What I failed to notice was that the clasp was not only old but also complicated; fastening it was a multistep process that I neglected to complete. All seemed well until I had one hand on the Bible and the other in the air. Then, a glance down revealed the pin hanging sideways. With all the hubbub, I had no time to correct the problem until after most of the photos were taken, showing my beautiful pin only in profile, contributing nothing to the symbolism of the moment but much to my angst. Years later, when publishing my memoirs, I tried to make amends by wearing the eagle—properly fastened—on the cover.

By tradition, it is the vice president, not the president, who administers the oath of office to a cabinet member. Here, the president looks on as Al Gore and I get in some practice just outside the Oval Office. At that point, my eagle pin was still secure.

Flanked by President Clinton and Vice President Gore,
I deliver remarks following my swearing in. My precious
eagle is barely hanging on. *Secretary of State Diamond
Eagle, designer unknown.*

To Madeleine–who leads fearlessly where others may fear to tread– with great pride and affection from your friend in The "Girls Room"– Hillary 1996

During an overseas trip, I needed to confer privately with First Lady Hillary Rodham Clinton. Where better than the ladies' room? I was proud to be the first woman to serve as secretary of state and delighted when Secretary Clinton became one of my successors. Opposite is a pin showing the glass ceiling in its ideal condition: shattered.
Breaking the Glass Ceiling, Vivian Shimoyama.

III. Body Language

By the time, in early 1997, when I began serving as secretary of state, my penchant for pins had become well-known. It helped that the picture on the front of *Newsweek* featured me with my combination Uncle Sam's hat and eagle. Since I was wearing brooches and getting photographed more than ever, the public's perception of the connection grew. Due to the demands on my time, I had fewer opportunities for browsing through shops, but it didn't matter, because everyone began giving me pins.

Newsweek cover, February 10, 1997. Photograph
by Timothy Greenfield-Sanders.

When diplomats meet, it is considered only civilized to
exchange gifts. Legally, American officials may retain foreign offerings that are below a certain value—in my day, $245. More expen-
sive items become the property of the U.S. government and are
displayed, stored, or sold for the benefit of the federal treasury.
Another option is to purchase the present at full price, which I did
on a few occasions. Some particularly large gifts, such as the handsome live horse with which I was presented in Mongolia or the
endearingly vocal goat I was given in Mali, are actually retained by
the hosts and, I suspect, given more than once to dignitaries
passing through Ulan Bator or Bamako.

Selecting the perfect gift for a foreign minister is like finding
"just the right thing" for a distant relative. The choice requires a
blend of common sense, intuition, and guesswork. I generally gave
mementos that reflected the United States: to men, eagle cuff links;
to women, a specially made eagle pin that I signed on the back.

Page 76: Atlas, Hervé van der Straeten.

My gifts to foreign leaders. *Foreign Minister's eagle, Christine Harkins; eagle cuff links, Ann Hand.*

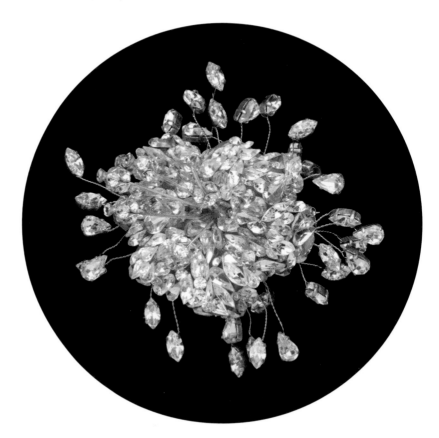

Clockwise from top left: Solana's flower, Primakov's snowy scene, Védrine's French design, designers unknown; Axworthy's maple leaf, Ann Hand.

Robin Cook gave me this striking Judith Leiber lion pin. I made sure to wear it during our press conference in 2000.

Scripture instructs us that it is more blessed to give than to receive, but it says nothing about which is more fun. My colleagues in the diplomatic community were pleased to assume that, in my case, a clever but inexpensive pin would always be appreciated. They were right. From British Foreign Secretary Robin Cook, I was given a lion brooch; from Canada's Lloyd Axworthy, a maple leaf; from France's Hubert Védrine, a sparkly French design; from NATO's Javier Solana, a delicate flower; and from Russia's Yevgeny Primakov and Igor Ivanov, lacquer pins showing various snowy scenes hand-painted in the intricate Russian style. You might think that enough would be enough, but to an aspiring collector, every addition is exciting. When presented with a gift-wrapped box, I ripped the ribbons off with heartfelt thanks and relish. The only problem I had was remembering to wear the pin in my next meeting with the person who had given it. As my Wellesley classmate Judith Martin (Miss Manners) might have reminded me, etiquette counts.

Among my favorite gifts is one from Leah Rabin, the widow of Israeli Prime Minister Yitzhak Rabin. The pin is of a dove, symbolizing the goal—peace in the Holy Land—for which the

In September 1999, Egyptian President Hosni Mubarak (*second from left*) and I witnessed the signing of an interim agreement between Israeli Prime Minister Ehud Barak and Palestinian Chairman Yasser Arafat.

prime minister had given his life. Like many of my predecessors, I had been reluctant to wander into the quicksand of Middle East negotiations. A series of terrorist incidents in the summer of my first year as secretary, however, left me with no choice. If leaders did not find a way to bring people together, extremists on every side would prepare for a future without peace, pointing inevitably to disaster.

In 1997, on August 6, I appeared before the National Press Club to outline ideas for negotiation and to announce plans for a trip to the region. The speech drew a full house, which, when combined with the television lights, warmed the room. I felt flushed and would probably have fainted had I not been petrified

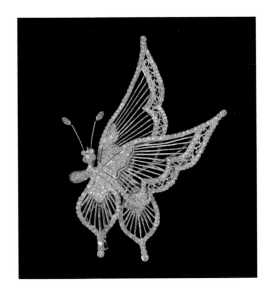

by what the newspapers would have written. Somehow I made it through the speech; Leah Rabin, among others, noted the dove pin displayed prominently on my chest.

A few weeks later, Mrs. Rabin came to see me at my hotel in Israel. She brought with her a companion necklace, composed of a flock of doves, and handed me a note that read: "There is a saying: 'One swallow doesn't announce the spring'—so maybe one dove needs reinforcements to create a reality of peace in the Middle East. We need hope which is so much lost—I do wish you will restore it. With all my sincerest wishes, Leah."

Above: A gift from Chairman Arafat. *Butterfly, designer unknown.*

Right: Dinner with Israeli Prime Minister Rabin, whose assassination in 1995 was a profound tragedy.

Below: Speaking on Middle East peace at the National Press Club. My dove was flying, but I felt faint.

I wore the dove pin again when paying my respects to the victims of genocide in Rwanda, 1997. *Peace dove and necklace, Cécile et Jeanne.*

Diplomatic negotiations often proceeded more slowly than hoped. I stocked up on turtles to signify my impatience and wore the crab when aggravated. *Crab, Vertige.*

Opposite page: Black and white turtle, Lea Stein; two purple, black, and gold turtles, Isabel Canovas; other designers unknown.

In the three years that followed, I devoted more time to the Middle East than to any other region, as did President Clinton. Although I often wore the dove, I found cause—when displeased with the pace of negotiations—to substitute a turtle, a snail, or, when truly aggravated, a crab. Sadly, none of the pins proved equal to their assigned task. Today, long after Mrs. Rabin's hope-filled gesture, the dove remains in need of reinforcements.

The frustrations of Middle East diplomacy were a constant reminder of the responsibilities that come with the job of secretary of state. I loved representing the United States but never stopped wondering how well I would measure up; thus I never stopped working. This attitude was reflected in a pin I had bought in Paris, made of gilt metal and wrought into a stylized Atlas holding up the Earth. I felt that America's duty was not to try to do everything itself, but to foster a sense of commitment that would bring out the best in every country. My intent in wearing the pin—which I took only to the most important meetings—was to indicate to my colleagues that, collectively, we had the weight of the world on our shoulders. As a joke, my diplomatic security team made up a T-shirt that portrayed me as Atlas, a role with which I would have been uncomfortable for two reasons: First, in most early depictions, Atlas appears naked; second, his actual task in Greek mythology was not to hold up the Earth—which was considered flat—but to hold up the heavens. Although my spirit would have been willing, I am much too short for that.

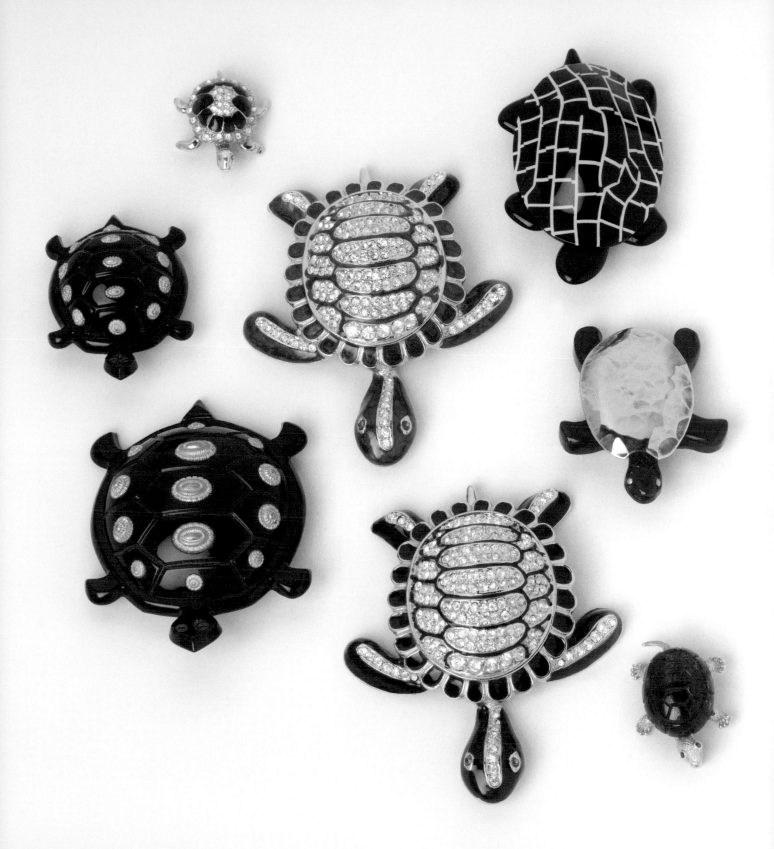

Green and red balloons, Swarovski.

As more people began to comment on my pins, I naturally found myself growing self-conscious. In the morning or even the night before, I started thinking about the right pin for the coming day and sometimes for each meeting. I didn't have much leisure for planning trips abroad, so often I just scooped up a handful of pieces from my jewelry box in hopes of finding an appropriate choice when the moment arrived. Some pins were essentially mood pieces, to indicate whether events were going poorly or well. When feeling good, I often wore a ladybug pin, because who doesn't love a ladybug? A second preference was my hot-air balloons, which I interpreted to mean high hopes, not overheated rhetoric. Other pins were aimed at conjuring up the quality needed to make a negotiation succeed, such as a tranquil swan or a wise owl. Less imaginatively, when discussing the salmon industry with my Canadian colleagues, I wore a pin shaped like a fish.

Wise owl, Lea Stein.

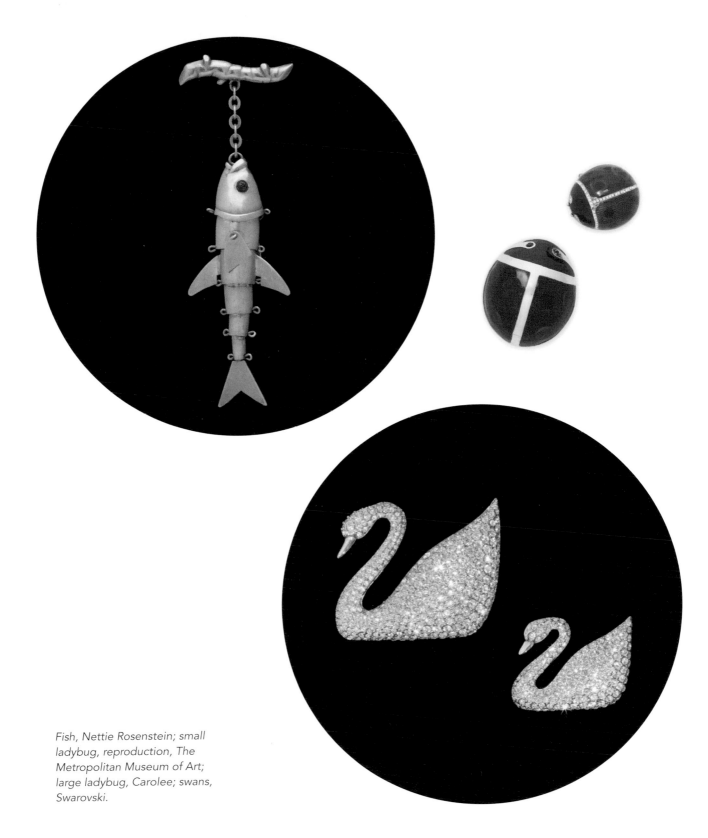

Fish, Nettie Rosenstein; small
ladybug, reproduction, The
Metropolitan Museum of Art;
large ladybug, Carolee; swans,
Swarovski.

In 1998, terrorists bombed the U.S. embassies in Kenya and Tanzania. Before flying across the Atlantic to honor those who were killed, I made brief remarks at Andrews Air Force Base. With sadness in my heart, I turned for help to an angel. *Angel, designer unknown.*

Left: David Yurman created this American flag pin in support of families affected by 9/11.

Sunburst, Hervé van der Straeten.

Because I am by nature a worried optimist (as opposed to a contented pessimist), I found many opportunities to wear my brooch of a brilliantly shining sun. Of course, part of being a diplomat is to make the best of a difficult situation, so I sometimes wore the sun more as an expression of hope than of expectation. In Haiti, for example, the Clinton administration had used force to oust an illegitimate military junta and restore the elected president. On every visit thereafter, I met with the civilian leaders, voiced America's desire to help, and talked about the prospects for progress. Each evening, as I put away the sun, I feared that neither my words of hope nor my effort to suggest the start of a new day would be enough to transform a desperate reality. The Haitian people—anxious and impoverished—deserve a far better government than they have had.

Naturally, not every diplomatic encounter demands a sunny attitude. If I wanted to deliver a sharp message, I often wore a bee. Muhammad Ali used to boast that he would "float like a butterfly, sting like a bee"; my message was that America would try to

With *Marine One* in the background, President Clinton by my side, and the sun on my shoulder. Latin America, 1999.

resolve every controversy peacefully, but if pushed into a corner, we had both the will and a way to strike back.

All this, I subsequently found, was generating not only diplomatic sparks but also an economic jump start for the costume jewelry industry. In Paris, I went into the gallery where Leah Rabin had bought her dove and was startled to find my picture on the wall. Visiting an antique jewelry store in New York, I was thanked by the owner for saving her business. A group in the Northeast set up a pin watch over the Internet where they would find out what I was wearing each day and try to interpret my choice.

Yasser Arafat and I conferring by phone with President Clinton. I spent many hours wrangling with the Palestinian leader about the need for compromise in the Middle East. My pin reflected my mood. *Bee, designer unknown.*

Feature articles appeared in the foreign and domestic press, and—to my embarrassment—total strangers began walking up and trying to give me pins.

One instance in particular bears recounting. On July 4, 2000, I was privileged to stand on the steps of the home of Thomas Jefferson, the first secretary of state, to witness hundreds of people take their oath of allegiance as new citizens of the United States. A naturalized citizen myself, I was moved by the ceremony and astonished, as always, by the remarkable diversity in background of the American people. Appropriately, attendees were given small American flags to wave, but at the accompanying reception my attention was drawn to a more dramatic version of the flag. Two elegant Virginia ladies, Julann Griffin and her sister, Maureen, were introduced to me, the former wearing a jumbo star-spangled U.S. flag brooch. When I complimented Julann, she offered me the pin. I had to say no but later accepted when she repeated her kind gesture after I left office. That was when Mrs. Griffin informed me that the brooch had originally been a gift from her ex-husband, the one and only Merv Griffin, who credited her with thinking up the game show *Jeopardy*. For me ever since, the question "What is Monticello?" has been linked to "The place where I got my fantastic pin."

Throughout my life with brooches, there has been one overriding challenge: how to wear them.

I long ago stopped wearing long necklaces, because they bounced around. I never liked the appearance of pins on lapels, especially on me. Pins on coats did not suit me either. I always preferred to wear brooches on the left side, thinking they looked better there, but the larger ones got in the way when I carried a purse with a strap. Smaller pins gradually appealed less because I had grown accustomed to what one reviewer of my books referred to as my "big honker brooches."

My Monticello flag, *Butler & Wilson*.

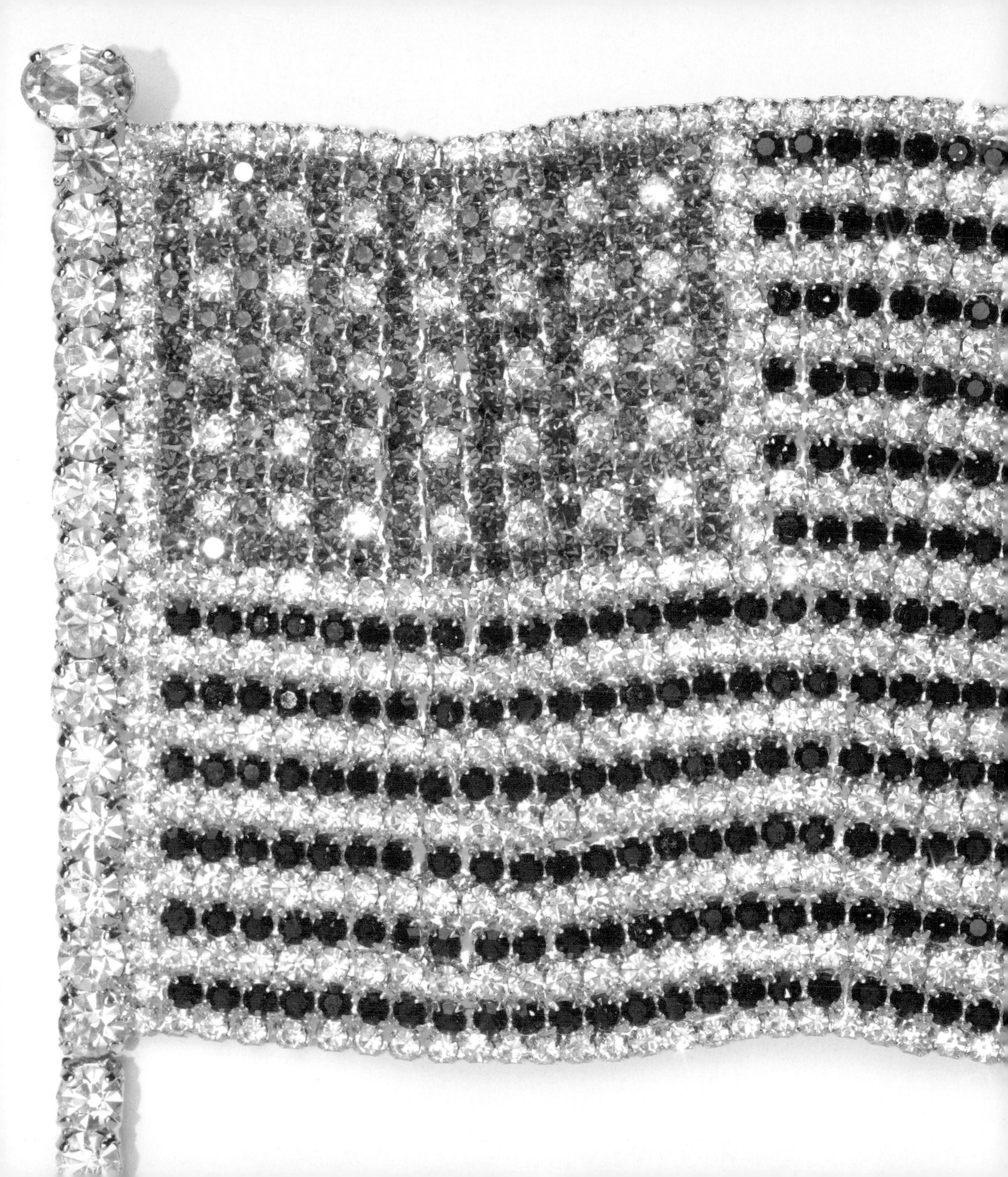

Hats, like jewelry, can be expressive. I spent my teenage years in Colorado and developed a fondness for Stetsons. I rekindled this affection when traveling as secretary of state, both because I liked the look and because I had more bad hair days than I could count. *Cowgirl hat, Ultra Craft.*

I like to arrange my bees and flower in different ways. One evening, while sitting with Aga Khan at a State Department dinner on cultural diplomacy, I had the bees in an ascending line.

Above: Spiders and their web, reproductions, The Metropolitan Museum of Art.
Opposite page: Flower with four bees, Joseff of Hollywood.

There was also the dilemma of how to arrange multiple pins. Some went together naturally, such as the zebras that I wore to a meeting with Nelson Mandela. Other combinations took more imagination—for example, bees approaching a sunflower. It was fun experimenting with various arrangements, but the practice threatened to consume too much energy. This is beside the fact that my clothes began to resemble dartboards, so perforated were they by the pins; eventually, I had to wear more and even bigger ones to mask the destruction.

At the same time, I had to deal with what a male friend described to me in jest as the Hooters issue. At Morey Junior High School in Denver, I had worn a blouse with a decoration that included two spiderwebs made of white stitching. I also owned a pair of small bug pins that I did not hesitate to place at the center of the spiderwebs, which were located on the left and right sides of my torso, face-front, breast-high. Given my age at the time, no one pointed out that they resembled pasties. With maturity comes growth, however, and as I traveled the world as a diplomat, I wanted people to look at my pins without embarrassing either the observers or me. So I wore the pins higher and higher up.

Nelson Mandela represented a new hope in Africa in the mid-1990s. I wore my favorite zebra pins when I met him at his estate in Pretoria, South Africa, in December 1997. *From left: Medium zebra, Ciner; large zebra, KUO; small zebra, designer unknown.*

I was justified in this approach when the foreign minister of South Korea made a comment, intended to be off the record, that he enjoyed hugging me at meetings and press conferences because I had "firm breasts." When the remark hit the newspapers, the foreign minister almost lost his job. Upon being asked to comment, I said, "Well, I have to have something to put these pins on." After that, the controversy quieted, but when I next met the foreign minister, instead of embracing we stopped an arm's length apart and shook hands.

One reason I had so many meetings with the foreign minister of South Korea is that we had so many quarrels with North Korea. That country's dictator, Kim Jong-il, had begun testing long-range missiles of a type that could conceivably threaten the territory of the United States. We were determined to prevent that, and so I traveled to Pyongyang, North Korea's capital, to negotiate.

In no other country on Earth are pins more crucial or less decorative. Every North Korean is expected to wear a pin bearing the image of the nation's founder, Kim Il-sung. Failure to display this badge of adoration is evidence of independent political thought, something strictly prohibited and severely punished. This is one reason why I find it absurd when U.S. politicians are criticized for not wearing American flag pins. The United States is a strong, confident country; we need not be so insecure as to require constant demonstrations of allegiance. At the same time, I wore the boldest American flag pin I had when meeting with Kim Jong-il. North Koreans are taught from an early age that America is evil; I wanted them to reconcile that reputation with photos of their exalted leader playing host to me.

Evil, of course, resides in the eye of the beholder. One of my more distinctive pieces of jewelry conveys a message about evil and how to resist it. The story begins in the spring of 1999, when leaders from NATO gathered in Washington to observe the

Foxy Lady, Lea Stein.

In North Korea, October 2000, posing for the cameras
with Kim Jong-il. To appear taller, I wore heels. So did he.
American flag, Robert Sorrell.

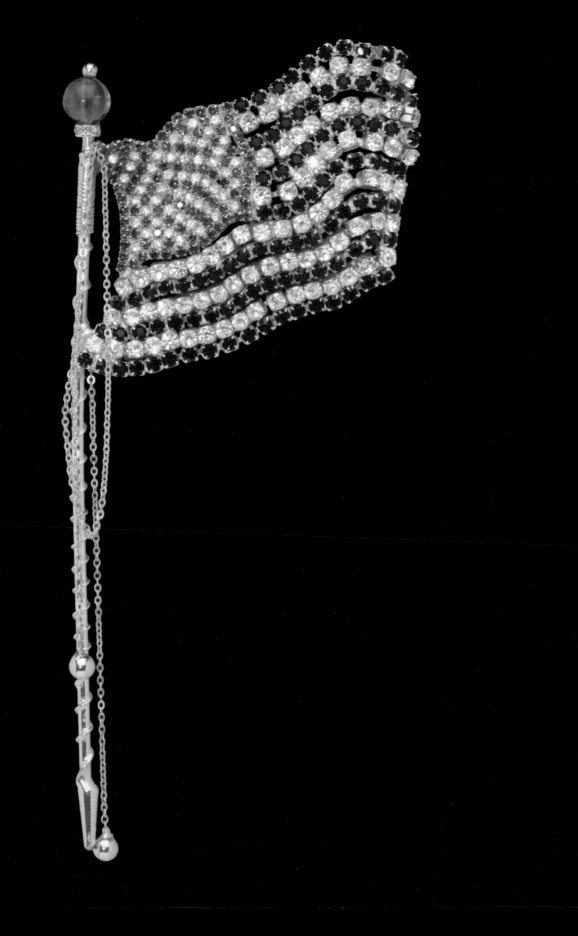

Clowning around with Defense Secretary Cohen and President Clinton.

alliance's fiftieth anniversary. As part of our preparations, President Clinton met with his foreign policy team. Just as we were getting down to business, photographer Diana Walker was allowed in. The photo op was good for public relations but meant that we had to cease talking about confidential issues. To dramatize the need for discretion, the president, clowning around, clamped his hand over his mouth. Defense Secretary Bill Cohen then put his hands over his ears. Taking my cue from the other two, I promptly covered my eyes. We literally made monkeys of ourselves before the camera, mimicking the well-known "Hear no evil, speak no evil, see no evil" adage.

At the time of the Walker photograph, I didn't own a three-monkey pin, but I soon found a set in Brussels. The individual figures are carved out of tagua nuts and each sits on a glass cabochon (pink, purple, or orange) encircled with crystals. The origin of the monkeys as a warning against temptation is lost in the mists of Japanese folklore, but the admonition dates back at least five hundred years and has much to do with accepting responsibility for wrongful thoughts and actions. The most famous carvings of Kikazaru (the "hear no" monkey), Iwazaru (the "speak no" monkey), and Mizaru (the "see no" monkey) can be found above the door of the seventeenth-century Toshogu shrine in Nikko, Japan.

Opposite page: Hear No Evil, Speak No Evil, See No Evil, Iradj Moini.

108

I first had occasion to wear the monkey pins on a visit to Moscow for a meeting with Russian President Vladimir Putin. One of the issues I wanted to raise was Russia's callous attitude toward human rights in the region of Chechnya, where brutal fighting was then taking place. The Russian military had legitimate reasons to fight rebel terrorists, but its approach was so heavy-handed that it was only creating more enemies. I argued that international monitors should be allowed into the region to protect civilians. Putin blocked the request, denying that any human rights violations were being committed. He saw no evil; hence my pins.

Despite our disagreements over Chechnya, the Russians were ever mindful of the signals I was sending. Putin told President Clinton that he routinely checked to see what brooch I was wearing and tried to decipher its meaning. Sometimes my choice reflected warmth in our relationship, as when I wore a gold spaceship brooch celebrating our partnership in the skies, but more often the mood was tense. Putin, who was young and disciplined, had replaced Boris Yeltsin, who was neither. My first impressions of the Russian leader were mixed—he was obviously capable, but his instincts appeared more autocratic than democratic. As the months passed, my early hopes were deflated by Putin's single-minded pursuit of power.

Among our most contentious discussions with the Kremlin were those involving nuclear arms. The United States wanted to make changes in the antiballistic missile treaty, and our counterparts did not. At the beginning of our talks, the Russian foreign minister looked at the arrow-like pin I had chosen for that day and inquired, "Is that one of your interceptor missiles?" I said, "Yes, and as you can see, we know how to make them very small. So you'd better be ready to negotiate."

As the debate about missiles showed, Cold War habits were slow to disappear. One December day in 1999, Stanislav Borisovich

One high point in U.S.-Russian relations occurred in 1998, when modules from our two countries linked up at the International Space Station. In Florida, I witnessed the night launch of the space shuttle *Endeavour*, which carried the U.S. module to its rendezvous. *Space shuttle pin, RC2, Corp.*

*Above: Interceptor Missile,
Lisa Vershbow.*

Left: With Russian foreign minister
Igor Ivanov on the balcony of the
State Department.

Gusev, a fiftyish "diplomat," was arrested while sitting on a bench outside the State Department. He was, in fact, a spy harvesting data from a listening device that our agents had located in a conference room at the far end of the building from my office. The electronic bug had been hard to find; Gusev had not. To avoid detection, the Russians had used a battery with low power, but this meant that anyone listening to the signals had to be stationed nearby. Gusev spent much of that autumn ostentatiously maneuvering his car outside the department's heavily guarded building, while inside our security people were scouring floors, walls, and furniture for whatever was prompting his movements.

Page 112: Sorcerer, Z. Alandia; other designers unknown.

Page 113: UFO, Jonette Jewelry.

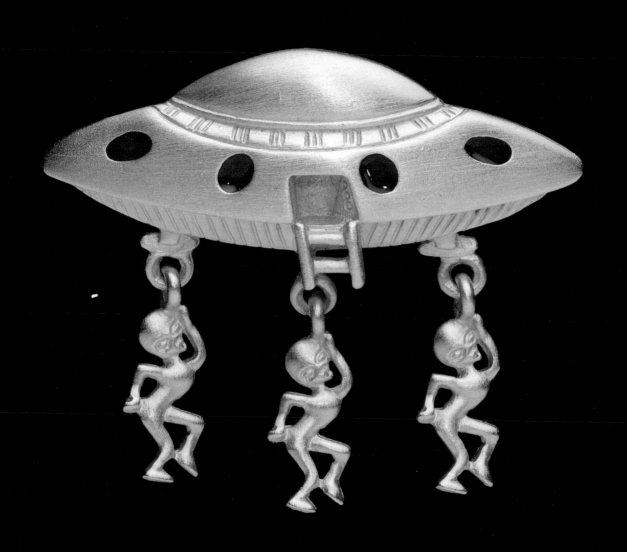

Above: Perhaps it is my imagination, but this pin always seems to end the day higher on my jacket than where it began.

Opposite page: Bug, Iradj Moini.

The incident attracted unwelcome publicity, but the Russians learned nothing from their eavesdropping that we wouldn't have told them if asked. Nor did the episode disrupt our diplomatic relations with Moscow, which have survived far more embarrassing cases of espionage. I met with Foreign Minister Ivanov in Europe only a few days subsequent to Gusev's arrest. We greeted each other as the friends we were, but Ivanov could not fail to notice on my dress a pin in the shape of an enormous bug.

I was reminded while secretary of state that there is a political dimension to the operations of the gem industry. Valuable resources attract feverish competition for access and control. To regulate the market, the world has created a system that encourages trade based on agreed-upon standards and rules. In some cases, as with endangered species, those rules prohibit trade. In others, our leaders have found it necessary to limit or ban sales from particular countries. Two examples during my tenure are worthy of mention.

Jade has been called the stone of Heaven. It is a personal favorite of mine and has been sought after for centuries, initially by Chinese emperors and Asian warlords, more recently by lovers of fine gems on every continent. Carat for carat, jade's value has soared. It is disquieting, then, that the majority of the world's most precious jade (or, more properly, jadeite) is mined in Burma, home to some of the poorest people and one of the most repressive governments on Earth. Until the mid-1990s, ethnic groups controlled the mines, using the revenue to preserve autonomy from the military regime. Over the past decade, the government has seized control of the mines, exploiting them (and the beaten-down souls who labor in them) for money and power. While in office, I championed economic sanctions against Burma; these have since been extended to include the most lucrative types of Burmese gems that are processed elsewhere. The ban is firmly supported by the Jewelers Vigilance Committee (a legal compliance group), the trade association Jewelers of America, and such leading international firms as Cartier and Tiffany.

In 1999, I visited a camp for amputees in Sierra Leone. It was a sweltering, muddy, crowded place. I remember especially holding a three-year-old girl who wore a red jumper and played with a toy car, using the only arm she had. Like many poor countries, Sierra Leone required voters to dip their fingers into indelible ink to prevent double-voting. The best-equipped rebel group felt it could frustrate the elections by chopping off the hands of potential voters, including children. This militia, and others in Angola and Congo, was financed in part by what came to be known as "blood" or "conflict" diamonds. These were diamonds seized and trafficked by armed groups that killed indiscriminately, often employing preteen soldiers.

Human rights activists appealed to me to try to stop the commercial use of such stones to fuel civil wars in Africa. I agreed. We supported a diplomatic initiative—known as the Kimberley Process—that is now accepted by every major diamond-producing and diamond-consuming country. Its purpose is to ensure that the much-coveted stones are traded legitimately from the time they leave a mine until the moment they appear in storefront windows. Like any such system, it is not leakproof, but it has done much to squeeze the profit out of blood diamonds, in part because the process has been widely backed by legitimate dealers. No responsible company wants to contribute to the success of thugs who start wars out of greed and hack off the limbs of children.

As a matter of policy, this story has an encouraging ending. On a personal level, it is even better. In 2007, I learned that the little girl in the red jumper whom I had tried to comfort in Sierra Leone had found adoptive parents and is now a happy and healthy teenager living on the same street as I do in Washington, D.C.

Panther, Cartier.

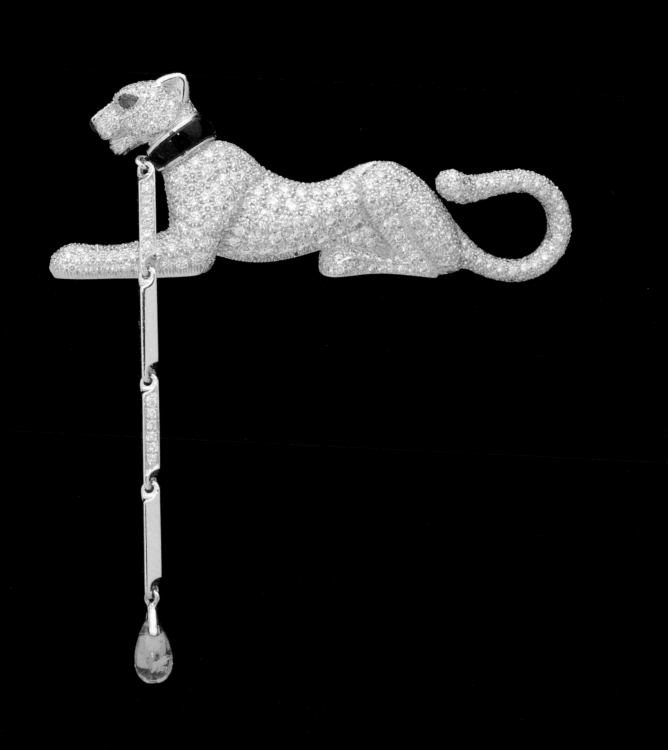

IV. "It Would Be an Honor"

The twentieth of January 2001 was my final day as secretary of state. I imagined that the incoming staff might have to drag me out of my office by the heels, but in the end I went peacefully. I had had my time; now it was the turn of others. That is how democracy works.

In my new life, I have worn many hats—as author, professor, speaker, and businesswoman. I serve as chair of the National Democratic Institute and president of the Truman Scholarship Foundation and have led task forces on poverty, genocide, and Arab democracy. World affairs remain my preoccupation, which means I continue to crisscross the globe. I also enjoy, now more than ever, wearing and collecting pins.

Above: I wore the Trailing Eagle pin
for our official cabinet photo in 2000,
my last full year as secretary of state.

Page 118: Trailing Eagle, Les Bernard.

This pen and book pin was a gift from my sister, Kathy Silva, upon the completion of my memoir, *Madam Secretary. Fountain pen, Carolee; book, designer unknown.*

Gift from the Harry S. Truman Presidential Library.

In Las Vegas years ago, I was booked to give a speech to a gathering of executives from the travel industry. The woman who organized the event asked what pin I intended to wear. I replied that I had brought only a necklace. She was aghast: "But that's impossible; we all expect you to wear a pin." Hours remained before the speech, and Las Vegas shops are always open, so I had little trouble finding something suitable. Since that time, I have learned to accept that when I appear in public, a pin is part of the package.

Fame, of course, is relative. In recent years, I have been mistaken in one venue or another for Margaret Thatcher, Barbara Bush, Judi Dench, Helen Thomas, some nice young fellow's Aunt Agatha, and the television weather lady in Minneapolis. Confusing my face with that of someone else is — in my ledger — a misdemeanor. Ignorance of my pins, however, is a felony. Among former foreign ministers, one of my closest friends is Joschka Fischer of Germany. After I left office, I was interviewed with Joschka on Berlin television. The commentator asked him what he thought of my practice of using pins to send a diplomatic message. Fischer hadn't a clue. He looked at me, then at her, then back at me, and confessed he couldn't have an opinion about something he had never noticed. No doubt Joschka will be delighted this Christmas to find under his *Weihnachtsbaum* an autographed copy of this book.

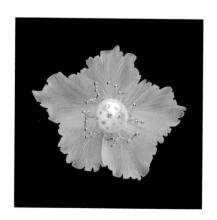

The 2008 election was one of the most exciting in memory. I was honored during the campaign to spend time working on foreign policy with the winner, President Barack Obama. *Left: Flower with pearl, Russell Trusso. Opposite page: Obama pins, Ann Hand.*

Arab village. The golden palm tree is from Saudi Arabia, the houses are from Egypt. *Palm tree, WRA; Houses of the Nile, Azza Fahmy.*

Opposite page: Kangaroos and hippo (with friend), St. John Knits.

Early in my life, my mother's ring served as a means for connecting one generation to another. When I was a young woman, the gift of a fraternity pin was an emblem of romance. In maturity, the brooches I bought for myself were signs of growing confidence and independence. In government, I used pins as a diplomatic tool. Now that I am out of office, my hobby often serves as an icebreaker. Before or after a speech, or while standing in line at the airport or supermarket, I am frequently asked about the pin I am wearing or to comment on one worn by somebody else. Such conversations, once initiated, can lead anywhere. I will not forget the woman who spoke enthusiastically about my pins before proceeding cheerily to compliment my overall appearance. "You look great," she said. "Just like my grandmother. She's 106 and as fit and sharp as she can be."

Although I remain busy, I do have more time than previously to shop in Washington and around the United States. I also often pick up pieces while overseas. In a less troubled world, we would

Speaking in San Diego at the Gemological Institute of America's fourth International Symposium, 2006. Given the nature of the event, I chose a particularly dramatic pin. *Opposite page: Dragon and sword, designer unknown.*

ordinarily think of jewelry as sending a friendly message, or at least not a violent one. In the post-9/11 era, however, even bottles of mouthwash and tubes of toothpaste can be considered threats. Perhaps I should not have been surprised, then, when a security agent stopped me at an airport gate and asked to examine a brooch I had just purchased in Turkey. The pin is of a slithery dragon wrapped around a small silver sword. Nothing to worry about, except that the sword is removable. The security agent glanced at me, then peered at the pin while shaking his head. "No weapons," he said.

Pages 128–129, row 1, from left: Sea horse, Swarovski; two colorful fish, Swarovski; rainbow fish, Swarovski; sea creature, Cécile et Jeanne; row 2: coral reef, designer unknown; sand dollar, designer unknown; lobster, Landau; row 3: crayfish, designer unknown; starfish, José & María Barrera; sea sponge, R. DeRosa; sea anemone, Ann Hand; octopus on coral, Kenneth Jay Lane; chambered nautilus, designer unknown.

Processing to Palm Sunday services in Tobago, April 1998. I didn't have a donkey or burro pin for the occasion, so I wore my circular horse.

I used to think rings were not worth buying because people have only ten fingers; I have to admit now to possessing more pins than any human could reasonably have occasion to wear. Of these, I bought many, but a goodly number more were gifts. Like the emperors of old India, I have become a collector and hoarder; still, most of my pins remain of the costume variety, hardly suitable for a royal procession. As is typical with a collector, I am attracted both to similarity and to diversity. It is always interesting to find a piece that is different from any other but also fun adding to categories I already have.

Opposite page: Leopard head, Ciner; teddy bear, Carolee; other designers unknown.

Nothing inspired me more than visiting American troops overseas. This pin was a gift from Barbara and Bill Richardson during his time as U.S. Ambassador to the United Nations. *Celebration of Freedom, designer unknown.*

Pin embossed with the Seal of the President of the United States, the White House. President Clinton's signature is on the back.

The center of my collection remains the Americana group, which has been filled out with flags, bows, ribbons, freedom torches, and even brooch-size replicas of the Statue of Liberty. One standout is a pin given to me by President and Mrs. Clinton that depicts the Seal of the President of the United States; another is a composition of emblems representing our various armed forces, accented with sparkling crystals and topped by an enameled American flag. This was a gift from Mary Jo Myers, whose husband, General Richard Myers, was my military adviser at the time and later chairman of the Joint Chiefs of Staff.

Opposite page: Ode to U.S. Armed Forces, Mina Lyles.

The Thelonious Monk Institute of Jazz celebrates one of America's distinctive art forms by educating young people and sending musical ambassadors around the world. Colin Powell and I served as cochairs for the Institute's twentieth anniversary celebration, at which Quincy Jones (*left*) and Herbie Hancock (*center*) presented a lifetime achievement award to Stevie Wonder. It's hard to tell from the picture, but I managed to get an entire jazz band onto my jacket. *Opposite page, amber musical instruments, Keith Lipert Gallery; other designers unknown.*

Eagle, Joseff of Hollywood.

Over the years, enough eagles have flown into my collection to comprise a small flock. One of the most interesting was produced by Joseff of Hollywood, who was famed for designing the jewelry in such films as *Gone With the Wind*, *The Wizard of Oz*, and the 1938 version of *Marie Antoinette*. Unlike the official U.S. eagle, this golden specimen clutches an olive branch on each side; its breast bears a shield garnished with red, white, and blue stones.

My most ingenious piece of Americana is a contemporary silver Liberty brooch. It shows the head of Lady Liberty, her eyes formed by two watch faces, one of which is upside down. The idea is that I can look down at the brooch to see when it is time for an appointment to end, while my visitor can look across at the pin for the same purpose. Dating from 1997, this design was made for "Brooching It Diplomatically," an exhibition inspired by my pins and organized by Helen W. Drutt English. The event drew contributions from dozens of international artists who were invited to create pins that transmitted a message, often a moral lesson about peace, justice, human rights, or some other uplifting goal.

Opposite page: Santa Fe eagles, Carol Sarkisian.

Helen W. Drutt English, an authority on modern and contemporary crafts and a curatorial consultant, was intrigued when she read that I had an unusual strategy for sending a political message. She invited jewelers from around the world to create pins that would send messages of their own. More than sixty artists from sixteen countries responded. Their imaginative contributions were displayed in "Brooching It Diplomatically: A Tribute to Madeleine Albright." This unique exhibit opened in Philadelphia, toured Europe, and was hosted by the Museum of Arts and Design in New York in 1999. Two of the pieces are now in my collection. On this page is an untitled leaf, which Helen Shirk, an American, created to illustrate the organic nature of negotiations. Opposite is Liberty, a pin designed by Gijs Bakker of the Netherlands. The clocks are arranged so that I, looking down, and a visitor, looking across, will each be able to tell when the time for our meeting is up.

Above: Meeting the press in Damascus, 1999. Is Syria ready to make peace? Not yet.

Right: This snake is far more beautiful than anything that slithers through the gardens of my farm in Virginia. *Snake, Kenneth Jay Lane.*

While in government, I thought first when selecting a pin about the utility it might have in diplomacy. This is because some figures are laden with meaning. The lion, for example, has been linked to power and the sun since the days of ancient Greece. Thus, Syria's formidable President Hafez al-Assad took considerable pride in the fact that his name means "lion" in Arabic. For our first meeting, I wore a lion pin, thinking it might put Assad in a forthcoming mood; it didn't.

The serpent, connected in my mind to Saddam Hussein, is often portrayed alongside a tree or, as on my pin, a branch. Together, the serpent and tree are considered symbols of life, fertility, and (because the snake sheds its skin) renewal and rebirth. The association has cultural and religious connotations dating all the way back

Opposite page: Lion, Kenneth Jay Lane.

Left: Two chicks, Tiffany & Co.

Opposite page: The dragonfly is an extraordinary species, with large eyes, two sets of powerful wings, an athletic body, and a healthy appetite for mosquitoes (center pin) and other pests. Known to the English as the "devil's darning needle," the insect is associated by the Japanese with courage, happiness, and strength. Artists find dragonflies fascinating; so do I. En tremblant dragonfly with pearl, Heidi Daus; turquoise enamel dragonfly, Ciner; yellow dragonfly, Swarovski; silver mosquito, Susan Sanders; other designers unknown.

to the Garden of Eden—the concept of which can be traced to Mesopotamia, or modern-day Iraq.

The dragon, meanwhile, has long symbolized China, as the bear has Russia, the koala Australia, and the mighty kiwi New Zealand. The Andean region is proud of its condor, Arabs of the peregrine falcon, Guatemala of the resplendent quetzal, Belize of the toucan, and the Bahamas of its flamingo. The United States may have a patent on the bald eagle, but other eagle species are claimed by a dozen lands, including Mexico's golden variety, Poland's white-tailed, Panama's harpy, and the African fish eagle of Zimbabwe and Zambia. The ubiquity of the great bird of prey is what prompted Benjamin Franklin to suggest for America a national symbol all its own: the turkey.

Out of government, I have less need to concern myself with such associations. I am free instead to indulge my own preferences, which include, in addition to patriotic symbols, such intriguing creatures as butterflies, frogs, songbirds, winged insects, and an infinite variety of bugs—especially big ones, the kind that seem poised to leap from my jacket. As with the lion and serpent, many of these species come with a past.

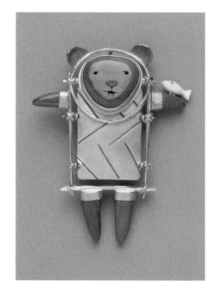

Shaman Bear, Carolyn Morris Bach.

The frog, for instance, is associated in many cultures with the creation myth, although I think of it more in the context of Moses and the second plague of Egypt ("The Nile will teem with frogs. They will come up into your palace and your bedroom and onto your beds."). As for the spider, it has been renowned since ancient times for its patience, wile, and predatory attitude. I wear my spider pin—complete with web and fly—when I am feeling devious; if you see it on any day except Halloween, beware.

The butterfly, emerging from the chrysalis, was considered by the Greeks to be a symbol of the soul. In the art nouveau period, around the end of the nineteenth century, a popular jewelry design showed the body of a woman with the wings of a butterfly. This symbolized the liberation of women.

The liberation of a country was commemorated by the Cartier company when, in 1944, Nazi occupiers were driven from France. The brooch showed an open cage and a bird singing. Two years earlier, when the storm troopers seized Paris, the company had produced a similar piece, except with the birds locked up.

Above: *L'Oiseau Libéré*, 1944, courtesy of Cartier.

Left: Archival drawing of *Oiseaux en Cage*, courtesy of Cartier.

*Right: Rose de Noël,
Van Cleef & Arpels.*

*Below: Moonstone Dandelion Puff,
Mauboussin.*

Flowers, too, are abundant in my collection. Like animals and bugs, various species of flora have acquired a distinctive meaning in literature and lore. The pansy is supposed to indicate thoughtfulness; ivy signifies fidelity; the lotus and the orchid were representative of the supposedly lethargic East; and the forget-me-not is a plea, well, not to forget. Flowers are usually for happier times, yet the lily has mournful connotations as well.

It would be inaccurate to suggest that I spend my spare time carefully arranging my brooches according to their affinity for one another. I am conscious, though, of the varieties that I have collected and am pleased to add to certain groups on occasion. There are, however, some unusual pin ideas that must be considered on their own.

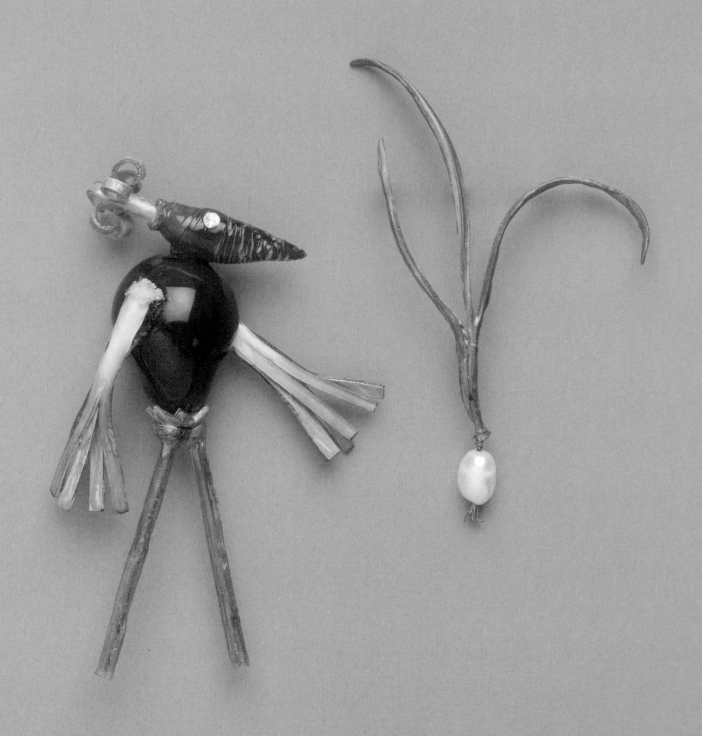

The smaller mushrooms represent Israel, Syria, and the Palestinian Authority; the larger is the United States. The pin was made from Syrian, Palestinian, Israeli, and American coins. *Mushrooms, Mary Ehlers.*

Page 150: Vegetable man, designer unknown; spring onion, Michael Michaud.
Page 151, row 1, from left: Apple, designer unknown; gold leaf with red berries, Cécile et Jeanne; row 2: Mount Vernon cherries, Michael Michaud/George Washington's Mount Vernon Ladies Association; red grapes, designer unknown; pomegranate, Cilça; row 3: three cherries, I. Chase; black cherries, Cilça; cluster of grapes, Bettina von Walhof.

Right: High-heeled shoe, designer unknown. Opposite page: Sailing ships, designer unknown.

On my sixty-fifth birthday, Elaine Shocas, my State Department chief of staff, gave me sixty-five pins, each costing less than three dollars. One of the gifts was in the shape of a high-heeled shoe. This was in commemoration of a comment I made when I was designated by Bill Clinton as the successor to Secretary of State Warren Christopher: "I only hope my heels can fill his shoes."

During Middle East peace talks, I was constantly besieged by the press. Journalists clamored to know everything about our meetings, even though the negotiators were pledged to secrecy. To deflect questions, I told reporters that peace talks were comparable to mushrooms, thriving only in the dark. My diplomatic security team soon surprised me with a custom-made pin depicting a tiny field of mushrooms. From then on, the mushrooms were a tip-off to the media that I had nothing revealing to say.

One set of pins that I bought for myself consists of a trio of brilliant enamels, each showing a ship at sea. Those familiar with history have asked me whether the ships represent the *Niña*, *Pinta*, and *Santa María*. I reply with a smile, for people should think what they want. In reality, I bought the pins with my three daughters in mind; the ships are beautiful, graceful, and moving along at full sail, having long since left home port.

I love spending time with children. Here, the Girl
Scouts are sporting merit badges; I'm wearing a fish.
Bejeweled Mickey, Disney Enterprises, Inc.

Speaking at the Smithsonian Institution National Museum of Natural History, Earth Day, 1998. On the opposite page is a group of environmental advocates.

Opposite page, row 1, from left: Grasshopper, Landau; cicada, Iradj Moini; row 2: fly with pearl, Iradj Moini, green ladybug, Sandor; row 3: two blue horseflies, designer unknown; row 4: green, purple, and blue beetle, Kenneth Jay Lane.

In 2008, I was invited to participate in an excursion to the Arctic along with an eclectic boatload of scientists, academics, businesspeople, philanthropists, musicians, and my grandson David. The sponsors were the National Geographic Society and the Aspen Institute. The theme was climate change; the scenery included melting ice and worried polar bears. Although others brought back photos and T-shirts, I returned with a pin. The gift of Stefan Rahmstorf, a professor of ocean physics, and his wife, Stefanie, a jewelry maker, the pin is shaped like a *C* with a white pearl attached at the top and bottom. The letter represents carbon; the round pearls are *O* for oxygen. Together, they symbolize carbon dioxide, a major cause of global warming. With each pin sold, the Rahmstorfs are able to buy and retire a ton of CO_2 from the European Union Emissions Trading System, thus reducing global emissions by that amount.

There is one other pin that is in a category by itself.

In the fall of 2006, I spoke at the D-Day Museum in New Orleans, at an event delayed for a year because of Hurricane Katrina. This gave me an opportunity to look around the city, large parts of which remained in ruins. I was saddened by the contrast between the museum—which celebrated America at its best—and the shabby treatment accorded to the residents of one of our country's most beautiful and historic cities.

At the reception following my speech, a young man bearing a small box approached me. Inside the box was a pin. "My mother loved you," he explained, "and she knew that you liked and wore pins. My father gave her this one for their fiftieth wedding anniversary. She died as a result of Katrina, and my father and I think she would have wanted you to have it. It would be an honor to her if you would accept it." I am not often speechless, nor am I quick to tear up, but this gift pushed me to the brink. The young man's father, I discovered, had earned two Purple Hearts fighting the

CO_2, Stefanie Rahmstorf; polar bear, Lea Stein.

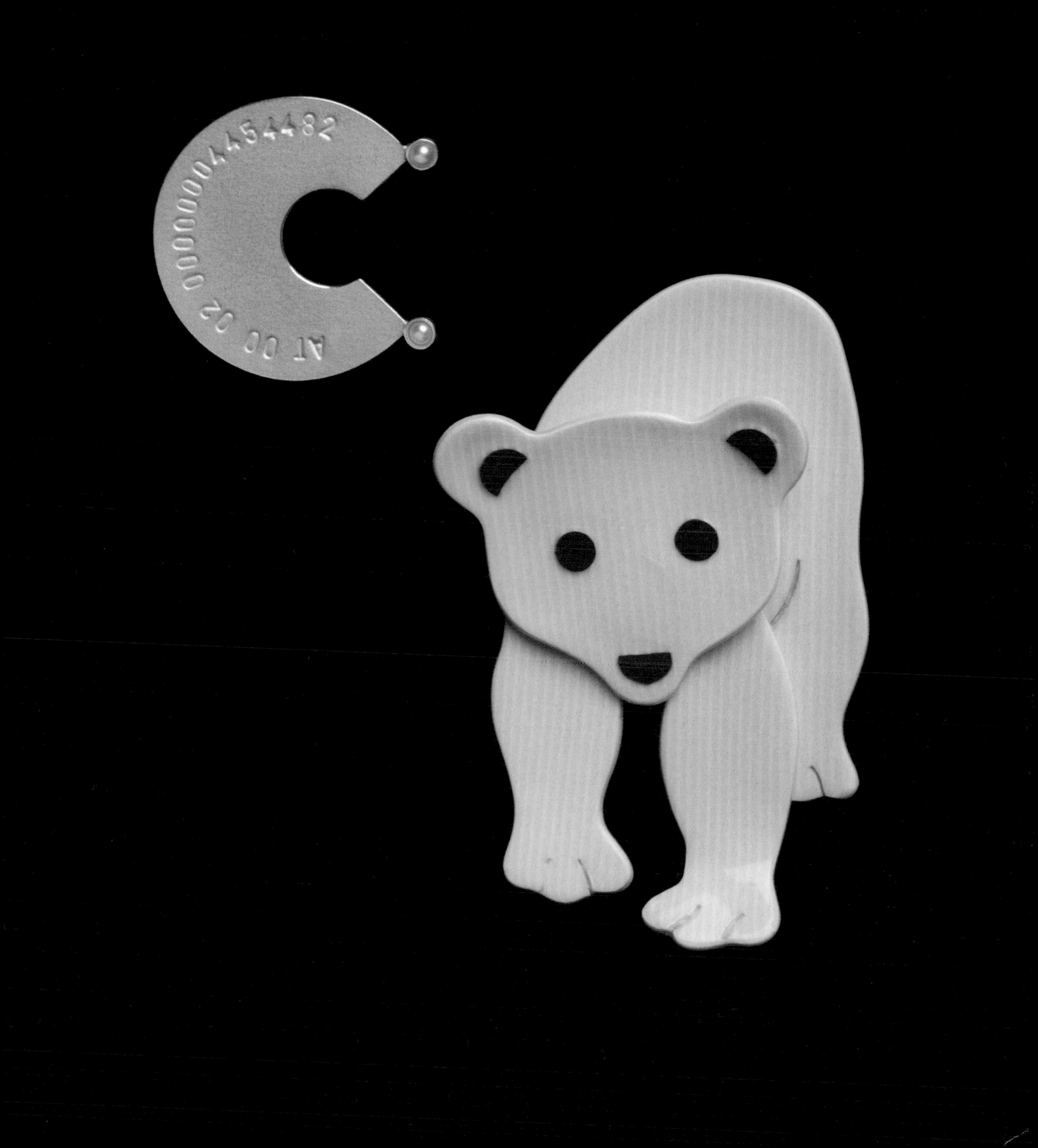
AT 00 02 0000004454482

Nazis in France, having suffered a bayonet wound and still carrying shrapnel in his left calf. His name is J.J. Witmeyer Jr.; he and his wife, Thais Audrey, were married for sixty-two years.

I call it the Katrina pin, a flower composed of amethysts and diamonds. I wear it as a reminder that jewelry's greatest value comes not from intrinsic materials or brilliant designs but from the emotions we invest. The most cherished attributes are not those that dazzle the eye but those that recall to the mind the face and spirit of a loved one.

Right: Wrapping Up Bow, designer unknown.

Opposite page: Katrina pin, designer unknown.

Pages 162–163, from left: Black-eyed Susan, Sandor; dandelion diamond puff and dandelion, McTeigue & McClelland; lily of the valley, designer unknown; tulip, designer unknown; Irish thorn, Michael Michaud; sunflower, Carolee; pearl flowers, JJ; gold and aqua flower, designer unknown.

As these pages illustrate, pins are inherently expressive. Elegant or plain, they reveal much about who we are and how we hope to be perceived. Styles have changed through the years, as has jewelry's role in relations between the genders and in the affairs of state. I was fortunate to serve at a time and in a place that allowed me to experiment by using pins to communicate a diplomatic message. One might scoff and say that my pins didn't exactly shake the world. To that I can reply only that shaking the world is precisely the opposite of what diplomats are placed on Earth to do.

PINDEX

 PAGE 1: THE GREAT SEAL OF THE UNITED STATES BOOK LOCKET AND PIN, 1990. ANN HAND, USA. 18KT YELLOW GOLD–PLATED BASE METAL. 1.4" X 1.4" (3.5CM X 3.5CM).

 PAGE 2: THE UNITED STATES CAPITOL, CIRCA 1970. MONET, USA. YELLOW GOLD–PLATED BASE METAL, RHINESTONES. 2" X 1.6" (5CM X 4CM).

 PAGE 4: ASYMMETRICAL GOLD HEART, 1996. ERWIN PEARL, USA. YELLOW GOLD–PLATED BASE METAL, GRANULATED FINISH. 2.8" X 2.2" (7.2CM X 5.6CM).

 PAGE 4: RED HEART AND BOW, 1996. ANN HAND, USA. YELLOW GOLD–PLATED BASE METAL, RHINESTONES. 2.5" X 1.3" (5.5CM X 3.5CM).

 PAGE 4: BEJEWELED HEART, CIRCA 1999. DESIGNER UNKNOWN, USA. OXIDIZED RHODIUM–FINISHED BASE METAL, RHINESTONES. 2.2" X 2" (5.6CM X 5CM).

 PAGE 4: SPARKLING RED HEART, 1998. ANN HAND, USA. YELLOW GOLD–PLATED BASE METAL, RHINESTONES. 1.8" X 2" (4.5CM X 5CM).

 PAGE 4: INTERLOCKING HEARTS, 1991. SWAROVSKI, AUSTRIA. YELLOW GOLD–PLATED BASE METAL, SWAROVSKI CRYSTALS. 2.8" X 1.9" (7CM X 4.8CM).

 PAGE 4: PURPLE HEART, CIRCA 1980. D.M. LEE, USA. STERLING SILVER WITH SUGILITE, AGATE, CORAL CABOCHONS. 1.3" X 1.2" (3.3CM X 3CM).

 PAGE 4: HAMMERED METAL HEART, CIRCA 2007. OMEGA, SWEDEN. HAMMERED AND OXIDIZED STERLING SILVER, GOLD. 4.1" X 1.5" (10.4CM X 3.9CM).

 PAGE 4: RHINESTONE BOMBÉ HEART, CIRCA 1997. DESIGNER UNKNOWN, ACQUIRED IN ARGENTINA. RHODIUM-PLATED BASE METAL, RHINESTONES. 2" X 1.8" (5.2CM X 4.5CM).

 PAGE 7: GOLD GINKGO LEAF, 2000. DESIGNER UNKNOWN, USA. GOLD-TONE BASE METAL, OXIDIZED STERLING SILVER, MARCASITES. 4.1" X 2.4" (10.4CM X 6.2CM).

 PAGE 7: SILVER GINKGO LEAF, 2000. DESIGNER UNKNOWN, USA. STERLING SILVER. 3" X 1.3" (7.5CM X 3.4CM).

 PAGE 7: COPPER GINKGO LEAF, 2000. DENNIS RAY/BEAUVOIR, THE NATIONAL CATHEDRAL ELEMENTARY SCHOOL, USA. COPPER-PLATED NATURAL LEAF. 2.9" X 2.8" (7.4CM X 7CM).

 PAGE 7: GOLD-STEMMED GINKGO LEAF, CIRCA 2000. FABRICE, FRANCE. GOLD-TONE BASE METAL, RESIN. 4.8" X 2.5" (12.3CM X 6.3CM).

 PAGE 9: VICTORY KNOT, 2008. VERDURA, USA. 18KT YELLOW-GOLD WIRE ROPE. 2.1" X 1.5" (5.3CM X 3.9CM).

 PAGE 11: ALERT LADY, 1999. BRIT SVENNI/BERIT KOWALSKI, NORWAY. OXIDIZED SILVER, SMOKY QUARTZ, BLACK BRUSH. 5.3" X 4.9" (13.4CM X 12.4CM).

 PAGE 12: BLACK RHINESTONE BUTTERFLY, 1999. ANN HAND, USA. BLACK-LACQUERED WHITE BASE METAL, RHINESTONES. 2.5" X 1.9" (6.3CM X 1.9CM).

 PAGE 12: GREEN AND CORAL BUTTERFLY, 2000. KENNETH JAY LANE, USA. YELLOW GOLD–PLATED BASE METAL, SIMULATED JADEITE, RHINESTONES, ENAMEL. 3.1" X 1.7" (7.8CM X 3.2CM).

 PAGE 12: BLUE BUTTERFLY, CIRCA 1950. DESIGNER UNKNOWN, FRANCE. MOLDED RESIN. 2.9" X 3.2" (7.3CM X 8.2CM).

 PAGE 12: LARGE SILVER BUTTERFLY, CIRCA 1990. © CHRISTIAN DIOR, FRANCE. TEXTURED, ENGRAVED SILVER BASE METAL, CRYSTALS. 4.4" X 3.5" (11.2CM X 8.8CM).

 PAGE 12: GOLD BUTTERFLY, CIRCA 1997. CÉCILE ET JEANNE, FRANCE. YELLOW GOLD–PLATED BASE METAL, CRYSTALS. 2.3" X 2.6" (5.8CM X 6.5CM).

 PAGE 12: GOLD BUTTERFLY AND WREATH, CIRCA 1994. MIRIAM HASKELL, USA. RUSSIAN GOLD–FINISHED BASE METAL, CRYSTAL ROSE MONTÉES. 2.1" X 2" (5.3CM X 5CM).

 PAGE 12: AMBER BUTTERFLY, CIRCA 1997. DESIGNER UNKNOWN, ACQUIRED IN LITHUANIA. YELLOW AMBER, WIRE. 2.6" X 2.0" (6.5CM X 5CM).

 PAGE 12: GREEN AND VIOLET BUTTERFLY, 2002. MODITAL BIJOUX, ITALY. YELLOW GOLD–PLATED BASE METAL, CRYSTALS. 1.8" X 1.6" (4.5CM X 1.6CM).

 PAGE 13: LIGHT BLUE RHINESTONE BUTTERFLY, CIRCA 1960. CINER, USA. OXIDIZED YELLOW BASE METAL, RHINESTONES, GLASS. 3.9" X 2.5" (10CM X 6.4CM).

PAGE 13: BLUE ENAMEL BUTTERFLY, CIRCA 1999. DESIGNER UNKNOWN, ACQUIRED IN HUNGARY. SILVER, ENAMELED SILVER, CRYSTALS, GLASS CABOCHONS. 2" X 1.2" (5.3CM X 3CM).

 PAGE 13: PEARL BUTTERFLY, 1997. KENNETH JAY LANE, USA. RHODIUM-PLATED BASE METAL, SIMULATED PEARLS, RHINESTONES. 3.6" X 2.6" (9.2CM X 6.6CM).

 PAGE 13: LAITICE FILIGREE BUTTERFLY, CIRCA 1997. CAVIAR, USA. STERLING SILVER, 18KT YELLOW GOLD. 3.1" X 2.5" (8CM X 6.4CM).

 PAGE 13: OPAL BUTTERFLY, 1999. TINY JEWEL BOX, USA. QUILPIE QUEENSLAND BOULDER OPALS FROM FLAME OPAL, AUSTRALIA, 18KT YELLOW GOLD, DIAMONDS, PLATINUM-WIRE ANTENNAE. 2.7" X 1.7" (6.8CM X 4.4CM).

 PAGE 13: RHINESTONE BUTTERFLY, 2000. JOSÉ & MARÍA BARRERA, USA. RHODIUM-PLATED BASE METAL, RHINESTONES. 2.5" X 1.7" (6.8CM X 4.4CM).

 PAGE 13: SILVER AND BLUE BUTTERFLY, CIRCA 1998. DESIGNER UNKNOWN, USA. STERLING SILVER, RHINESTONES. 2.3" X 2" (5.8CM X 5.1CM).

 PAGE 13: GRAY RHINESTONE BUTTERFLY, CIRCA 1960. CINER, USA. OXIDIZED WHITE BASE METAL, RHINESTONES. 3.1" X 1.8" (6.3CM X 4.7CM).

 PAGE 14: SERPENT, CIRCA 1860. DESIGNER UNKNOWN, USA. 18KT YELLOW GOLD, DIAMOND. 2.4" X 1.1" (6.1CM X 2.8CM).

 PAGE 18: BLUE BIRD, CIRCA 1880. ANTON LACHMANN, AUSTRIA. 14KT YELLOW GOLD, SILVER, ENAMEL, RUBIES, DIAMONDS. 4.3" X 2" (11CM X 5.1CM).

 PAGE 20: SUN, 2006. STEINMETZ DIAMONDS, SOUTH AFRICA. 18KT WHITE AND YELLOW GOLD, DIAMONDS. 1.4" X 0.4" (3.5CM X 1CM).

 PAGE 21: CRYSTAL FLY, 1998. © CHRISTIAN DIOR, FRANCE. RHODIUM-PLATED BASE METAL, CRYSTALS. 1.5" X 1.4" (3.8CM X 3.5CM).

 PAGE 21: GREEN AND BLUE RHINESTONE BEES, CIRCA 1960. CINER, USA. YELLOW GOLD–PLATED BASE METAL, RHINESTONES. 2.6" X 1.5" (6.5CM X 2.6CM).

 PAGE 21: TURQUOISE BEE, CIRCA 1940. WALTER LAMPL, USA. 14KT YELLOW GOLD, TURQUOISE, MOTHER-OF-PEARL. 1" X 1" (2.6CM X 2.5CM).

 PAGE 21: GOLDEN BEE, 1997. ST. JOHN KNITS, USA. YELLOW GOLD–PLATED BASE METAL, CRYSTALS. 2.3" X 0.8" (3.3CM X 2CM).

 PAGE 22: EYEGLASSES, CIRCA 2002. DESIGNER UNKNOWN, USA. YELLOW GOLD–PLATED BASE METAL, RHINESTONES. 1.7" X 0.7" (4.3CM X 1.9CM).

 PAGE 22: LIPS AND LIPSTICK, CIRCA 2002. DESIGNER UNKNOWN, USA. YELLOW GOLD–PLATED BASE METAL, ENAMEL, RHINESTONE. 1.8" X 1.9" (4.5CM X 1.9CM).

 PAGE 22: LEOPARD PRINT PURSE, CIRCA 1970. © AJMC, USA. ENAMELED SILVER-TONE BASE METAL, RHINESTONES. 1.2" X 1.2" (3CM X 3CM).

 PAGE 22: REPTILE PRINT PURSE, CIRCA 1970. © AJMC, USA. ENAMELED SILVER-TONE BASE METAL, RHINESTONES. 1.3" X 1.7" (3.2CM X 4.4CM).

 PAGE 22: RUBY SLIPPERS, CIRCA 2002. AJC, USA. YELLOW GOLD–PLATED BASE METAL, GLITTER. 2.3" X 1.5" (5.8CM X 3.8CM).

 PAGE 23: PETIT OISEAU, 1998. JACQUELINE LECARME, BELGIUM. RESIN, SIMULATED PEARL, SIMULATED HORN, CRYSTALS, BUTTONS, BEADS. 4.7" X 3.3" (12CM X 8.5CM).

 PAGE 24: LAUREL WREATH, CIRCA 1997. DESIGNER UNKNOWN, GREECE. STERLING SILVER, GOLD-FILLED. 2.8" X 3.1" (7.2CM X 7.8CM).

 PAGE 25: BYZANTINE SHIELD, 1970. ILIAS LALAOUNIS, GREECE. 18KT YELLOW GOLD, RUBY, SAPPHIRES, BAROQUE FRESHWATER CULTURED PEARLS. 2" X 2" (5CM X 5CM).

 PAGE 25: CIRCLE OF PEARLS, CIRCA 1990. CRAFT, USA YELLOW GOLD–PLATED BASE METAL, ENAMEL, GLASS CABOCHON, RHINESTONES, SIMULATED PEARLS. 2.2" X 2.2" (5.6CM X 5.6CM).

 PAGE 25: BIRD WITH PEARL, CIRCA 1990. BETTINA VON WALHOF, USA. BLACK RHODIUM–PLATED BASE METAL, BAROQUE SOUTH SEA CULTURED PEARL, RHINESTONES. 4.7" X 4.5" (12CM X 11.5CM).

 PAGE 26: CASTLE, CIRCA 2007. LJ, TURKEY. PLATINUM, DIAMONDS, PINK CHALCEDONY, CORAL, TURQUOISE, EMERALDS, MOTHER-OF-PEARL, LAPIS LAZULI, RUBIES, SAPPHIRES. 1.5" X 2.0" (3.7CM X 5.2CM).

 PAGE 27: INDIAN ELEPHANT, CIRCA 1960. DeNICOLA, USA. YELLOW GOLD–PLATED BASE METAL, RHINESTONES, GLASS CABOCHON. 2" X 2" (5.2CM X 5CM)

 PAGE 28: RHINESTONE FLEUR-DE-LIS, CIRCA 1993. DESIGNER UNKNOWN, USA. RHODIUM-PLATED BASE METAL, RHINESTONES. 2.1" X 3.4" (5.3CM X 8.6CM).

 PAGE 28: GOLD FLEUR-DE-LIS, CIRCA 1994. SOFIE, BOSNIA. 18KT YELLOW GOLD, STERLING SILVER, DIAMONDS, RUBIES, EMERALDS. 1.3" X 1.1" (3.3CM X 2.7CM).

 PAGE 30: WESTERN SUN, 2003. FEDERICO JIMENEZ, USA/ MEXICO. STERLING SILVER, HOWLITE, TURQUOISE. 3.7" X 3.1" (9.4CM X 3.1CM).

PAGE 31: EAGLE DANCER, CIRCA 1970. JERRY ROAN, USA. BOLO TIE MADE INTO A PIN. STERLING SILVER, TURQUOISE, CORAL. 6.9" X 5.2" (17.5CM X 13.2CM).

PAGE 32: MANDULA KORBEL'S THREE GOLD CIRCLES, CIRCA 1935. DESIGNER UNKNOWN, CZECHOSLOVAKIA. 18KT YELLOW GOLD. 1.7" X 0.9" (4.2CM X 2.2CM).

 PAGE 32: TITO'S RING, CIRCA 1946. DESIGNER UNKNOWN, YUGOSLAVIA. 14KT WHITE GOLD, EMERALD, DIAMONDS. 0.8" X 0.6" (2.0CM X 1.5CM).

 PAGE 33: MANDULA KORBEL'S SAPPHIRE, CIRCA 1945. DESIGNER UNKNOWN, CZECHOSLOVAKIA. 14KT PINK GOLD, SAPPHIRE, DIAMONDS. 2.6" X 1.3" (6.6CM X 3.4CM).

 PAGE 34: ORDER OF THE WHITE LION, 1997. OFFICE OF THE PRESIDENT, CZECH REPUBLIC. STERLING SILVER, ENAMEL. 3.6" X 3.6" (9.2CM X 9.2CM).

 PAGE 35: CZECH ART NOUVEAU DESIGN, CIRCA 1995. DESIGNER UNKNOWN (VM), CZECH REPUBLIC. STERLING SILVER, GARNET CABOCHONS. 1.7" X 1.6" (4.4CM X 4.2CM).

 PAGE 35: MUCHA-INSPIRED CORAL, CIRCA 2002. DESIGNER UNKNOWN, CZECH REPUBLIC. STERLING SILVER, CORAL, CRYSTALS. 2.2" X 2.1" (5.7CM X 5.3CM).

 PAGE 36: BIRD, 2000. IRADJ MOINI, USA. RHODIUM-PLATED ENAMELED BASE METAL, CRYSTALS, GLASS CABOCHONS. 5.2" X 3.5" (13.3CM X 9.2CM).

 PAGE 38: ALUMNAE LEAF, 1992. WELLESLEY COLLEGE, USA. 14KT YELLOW GOLD. 2.1" X 1.2" (5.3CM X 3CM).

 PAGE 39: FRATERNITY PIN, 1957. THETA DELTA XI FRATERNITY, USA. 10KT YELLOW GOLD, DIAMONDS, ENAMEL, SEED PEARLS. 0.7" X 0.6" (1.7CM X 1.5CM).

 PAGE 39: CIRCLE PIN, 1954. DESIGNER UNKNOWN, USA. 14KT YELLOW GOLD, ENAMEL. 0.6" X 0.6" (2CM X 2CM).

 PAGE 40: FEATHER, CIRCA 1940. DESIGNER UNKNOWN, USA. 14KT YELLOW GOLD, RUBIES. 2.4" X 1.0" (6.1CM X 2.6CM).

 PAGE 40: JADE DRAGON, CIRCA 1950. DESIGNER UNKNOWN, USA. 14KT YELLOW GOLD, JADEITE. 2.6" X 0.7" (6.5CM X 1.9CM).

 PAGE 41: RUBY FISH, 1959. DESIGNER UNKNOWN, USA. 18KT YELLOW GOLD, RUBIES, EMERALD. 1.6" X 1.9" (4CM X 4.8CM).

 PAGE 41: VIOLETS, CIRCA 1965. DESIGNER UNKNOWN, USA. 14KT YELLOW GOLD, AMETHYST, DIAMONDS, NEPHRITE JADE. 1.75" X 1.25" (4.3CM X 3.3CM).

 PAGE 42: BOHEMIAN GARNET SET, CIRCA 1860. DESIGNER UNKNOWN, BOHEMIA. 10KT ROSE GOLD, GARNETS. NECKLACE: 16.2" (41.1CM); BRACELET: 7.75" (20CM); PIN/PENDANT: 1.5" X 1.5" (3.8CM X 3.8CM); EARRINGS: 0.5" (1.3CM).

 PAGE 43: KATIE'S HEART, 1972. KATIE ALBRIGHT, USA. CLAY. 2.8" X 2.4" (7.2CM X 6CM).

 PAGE 45: POPPY, 1959. VERDURA, USA. 18KT YELLOW GOLD, PLATINUM, DIAMONDS, SAPPHIRE CABOCHONS. 2.6" X 1.7" (6.5CM X 4.4CM).

 PAGE 46: MELI MELO, 2001. ©CARTIER, FRANCE. 18KT WHITE GOLD, CHALCEDONY AND MOONSTONE CABOCHONS, PINK TOURMALINES, CORDIERITES, AQUAMARINES, DIAMONDS, MANDARIN GARNETS. 1.9" X 1.9" (4.8CM X 4.8CM).

 PAGE 47: WRAPPED HEART, 1946. VERDURA, USA. 14KT YELLOW GOLD, PINK TOURMALINES. 2.5" X 2.2" (6.4CM X 5.6CM).

 PAGE 48: SUFFRAGETTE PIN, CIRCA 1900. DESIGNER UNKNOWN, USA. YELLOW GOLD–PLATED MESH, GLASS, RHINESTONES, SIMULATED PEARLS. 2.6" X 2.4" (6.6CM X 6.1CM).

 PAGE 50: FRENCH URN, CIRCA 1900. DESIGNER UNKNOWN, FRANCE. SILVER, GLASS, CRYSTALS, SIMULATED LAPIS LAZULI, RESIN. 4.9" X 1.2" (12.5CM X 3CM).

 PAGE 51: SHEAF OF WHEAT, 1987. TIFFANY & CO., USA. 18KT YELLOW GOLD, DIAMONDS. 2.8" X 1.3" (7CM X 1.7CM).

 PAGE 52: GRAPES, 1995. TINY JEWEL BOX, USA. 18KT YELLOW GOLD, SOUTH SEA CULTURED PEARLS, DIAMONDS. 3" X 2" (7.5CM X 5CM).

 PAGE 54: BERLIN WALL, 1989. GISELA GEIGER, USA. SILVER, CONCRETE PIECE OF BERLIN WALL, SILVER WIRE. 2.9" X 2.8" (7.4CM X 7.2CM).

 PAGE 55: G-8 PIN, CIRCA 1997. DESIGNER UNKNOWN, MEXICO. STERLING SILVER. 2" X 1.8" (5CM X 4.5CM).

 PAGE 56: SAILOR, CIRCA 1940. MONET, USA. ENAMELED WHITE BASE METAL. 1.3" X 2.4" (3.3CM X 6.1CM).

 PAGE 57: MEMORIAL BOW, CIRCA 1945. TRIFARI, USA. RHODIUM-PLATED BASE METAL, RHINESTONES, ENAMEL. 2.4" X 2.2" (6CM X 5.5CM).

 PAGE 58: AMERICAN FLAG, 1996. ANN HAND, USA. YELLOW GOLD AND RHODIUM–PLATED BASE METAL, RHINESTONES. 2.1" X 1.7" (5.3CM X 4.4CM).

 PAGE 58: BRAVE HEART, 2001. SWAROVSKI, AUSTRIA. TRIBUTE TO THE VICTIMS OF 9/11. RHODIUM-PLATED BASE METAL, ENAMEL, SWAROVSKI CRYSTALS. 0.8" X 0.7" (2.9CM X 1.8CM).

 PAGE 58: AIDS RIBBON, CIRCA 2000. DESIGNER UNKNOWN, USA. YELLOW BASE METAL, CERAMIC. 0.9" X 1.2" (2.3CM X 3CM).

PAGE 58: HEART WITH DONKEY, 2004. DESIGNER UNKNOWN, USA. YELLOW GOLD–PLATED BASE METAL, ENAMEL. 1.6" X 1.4" (4CM X 3.5CM).

 PAGE 58: HEART STICKPIN, CIRCA 1995. REPRODUCTION, THE METROPOLITAN MUSEUM OF ART, USA. YELLOW GOLD–PLATED BASE METAL, ENAMEL. 2.6" X 0.7" (6.6CM X 1.9CM).

 PAGE 58: FRENCH RIBBON BOW, 1940. SILSON, USA. YELLOW GOLD–PLATED BASE METAL, ENAMEL. 2.8" X 1.6" (7.1CM X 4CM).

 PAGE 58: PATRIOTIC BOW, 1999. © CAROLEE, USA. YELLOW GOLD–PLATED BASE METAL, RHINESTONES, ENAMEL. 1.9" X 0.8" (4.7CM X 2CM).

 PAGE 58: STATUE OF LIBERTY, CIRCA 1940. DESIGNER UNKNOWN, USA. BAKELITE, SIMULATED MOTHER-OF-PEARL. 2.5" X 1.5" (6.4CM X 3.8CM).

 PAGE 58: SAFETY-PIN AMERICAN FLAG, CIRCA 1970. DESIGNER UNKNOWN, USA. BRASS, GLASS BEADS. 1.4" X 1.1" (3.5CM X 2CM).

 PAGE 58: SMALL AMERICAN FLAG, CIRCA 1960. DESIGNER UNKNOWN, USA. YELLOW GOLD–PLATED BASE METAL, RHINESTONES, WIRE. 1.7" X 1.1" (4.2CM X 2.8CM).

 PAGE 58: LARGE FIGHT FOR FREEDOM TORCH, CIRCA 1940. DMW, USA. YELLOW GOLD–PLATED BASE METAL, ENAMEL. 1.7" X 3" (4.2CM X 7.5CM).

 PAGE 58: WWII RIBBON, CIRCA 1940. SILSON, USA. YELLOW GOLD–PLATED BASE METAL, ENAMEL. 1.6" X 2.4" (4CM X 6.1CM).

 PAGE 58: STARS AND STRIPES, CIRCA 1960. CINER, USA. YELLOW GOLD–PLATED ENAMEL. 1.5" X 1.8" (3.7CM X 4.5CM).

 PAGE 58: SMALL FIGHT FOR FREEDOM TORCH, CIRCA 1940. DMW, USA. YELLOW GOLD–PLATED BASE METAL, ENAMEL. 0.7" X 1.1" (1.8CM X 2.9CM).

 PAGE 59: UNCLE SAM TOP HAT, CIRCA 1940. TRIFARI, USA. ENAMELED RHODIUM-PLATED BASE METAL, RHINESTONES. 1.5" X 1" (3.8CM X 2.5CM).

 PAGE 59: UNCLE SAM EAGLE, CIRCA 1940. TRIFARI, USA. ENAMELED RHODIUM-PLATED BASE METAL, RHINESTONES. 2.9" X 1" (7.3CM X 2.5CM).

 PAGE 60: CHINESE SHARD DRAGON, CIRCA 1997. DESIGNER UNKNOWN, CHINA. SILVER, PORCELAIN. 2.9" X 2.2" (7.4CM X 5.7CM).

 PAGE 61: COLORFUL BIRD, 1997. IRADJ MOINI, USA. YELLOW GOLD–PLATED BASE METAL, WIRE, EMERALD CABOCHON, RUBIES, AMETHYST, SIMULATED CORAL, COMPOSITE MALACHITE/AZURITE, CRYSTALS. 3.5" X 1.7" (9CM X 4.3CM).

 PAGE 63: ESM; SENATOR EDMUND S. MUSKIE, 1972. ©CARTIER, FRANCE/USA. 18KT YELLOW GOLD. 0.6" X 0.6" (1.5CM X 1.5CM).

 PAGE 63: SAXOPHONE, 1993. KENNETH JAY LANE, USA. YELLOW GOLD–PLATED BASE METAL. 3.1" X 1.2" (8CM X 3CM).

 PAGE 63: SOLIDARITY, CIRCA 1981. DESIGNER UNKNOWN, POLAND. WHITE BASE METAL, ENAMEL. 0.8" X 0.4" (2.1CM X 1.1CM).

 PAGE 63: RED DRESS FOR WOMEN'S HEART HEALTH, 2008. NATIONAL INSTITUTES OF HEALTH, USA. RHODIUM-PLATED BASE METAL, SWAROVSKI CRYSTALS. 1.3" X 0.7" (3.5CM X 2.2CM).

 PAGE 63: BREAST CANCER RIBBON, 2006. © SUSAN G. KOMEN FOR THE CURE, USA. PINK-ENAMELED RHODIUM–PLATED BASE METAL. 1.1" X 0.8" (2.9CM X 2CM).

 PAGE 63: TRUMAN CAMPAIGN BUTTON, 1948. DESIGNER UNKNOWN, USA. STEEL. 1.3" X 1.3" (3.3CM X 3.3CM).

 PAGE 63: KENNEDY CAMPAIGN BUTTON, 1960. DESIGNER UNKNOWN, USA. STEEL. 1.6" X 1.6" (4.5CM X 4.5CM).

 PAGE 63: JOHNSON CAMPAIGN BUTTON, 1964. DESIGNER UNKNOWN, USA. TIN. 2.2" X 2.2" (5.6CM X 5.6CM).

 PAGE 63: CARTER CAMPAIGN BUTTON, 1976. DESIGNER UNKNOWN, USA. TIN. 1.6" X 1.6" (4.4CM X 4.4CM).

 PAGE 63: CLINTON CAMPAIGN BUTTON, 1992. DESIGNER UNKNOWN, USA. ALUMINUM. 2.2" X 2.2" (5.6CM X 5.6CM).

 PAGE 63: OBAMA CAMPAIGN BUTTON, 2008. DESIGNER UNKNOWN, USA. ALUMINUM. 2.2" X 2.2" (5.6CM X 5.6CM).

 PAGE 65: PARTNERS IN PEACE, 1999. JANET LANGHART COHEN/ANN HAND, USA. YELLOW GOLD–PLATED BASE METAL, SIMULATED PEARLS, RHINESTONES. 4.7" X 2.9" (12CM X 7.3CM).

 PAGE 66: JAMBIYA DAGGER, CIRCA 1997. DESIGNER UNKNOWN, YEMEN. SILVER. 2" X 1" (5.2CM X 2.5CM).

 PAGE 66: ROCKET-PROPELLED GRENADE LAUNCHER (RPG), 1988. DESIGNER UNKNOWN, PAKISTAN. SILVER, LAPIS LAZULI. 2.7" X 0.8" (7CM X 2CM).

 PAGE 67: KNOT OF HERCULES, CIRCA 1970. ILIAS LALAOUNIS, GREECE. 18KT YELLOW GOLD. 2.8" X 0.6" (7.1CM X 1.6CM).

PAGE 68: LIBERTY EAGLE, 1992. ANN HAND, USA. 18KT YELLOW GOLD–PLATED STERLING SILVER, SIMULATED PEARL. 1.5" X 1.4" (4CM X 4CM).

 PAGE 73: SECRETARY OF STATE DIAMOND EAGLE, CIRCA 1890. DESIGNER UNKNOWN, FRANCE. 18KT YELLOW GOLD, SILVER, DIAMONDS, RUBIES, DROP NATURAL SALTWATER PEARL. 3.5" X 1.1" (8.8CM X 2.9CM).

 PAGE 75: BREAKING THE GLASS CEILING, 1992. VIVIAN SHIMOYAMA, USA. FUSED GLASS, 22KT GOLD TRIM. 3" X 1.6" X 2.6" (7.5CM X 4CM X 6.7CM).

 PAGE 76: ATLAS, 1991. HERVÉ VAN DER STRAETEN, FRANCE. GILDED, HAMMERED BRASS, GLASS CABOCHON. 1.9" X 5" (4.9CM X 12.7CM).

 PAGE 79: FOREIGN MINISTER'S EAGLE, 1997. CHRISTINE HARKINS, USA. MADELEINE ALBRIGHT'S SIGNATURE ENGRAVED ON BACK. YELLOW GOLD–PLATED BASE METAL. 2.6" X 1.4" (6.5CM X 3.5CM).

 PAGE 79: EAGLE CUFF LINKS, 1996. ANN HAND, USA. 18KT YELLOW GOLD–PLATED, STERLING SILVER, RUBIES. 0.7" X 0.6" (1.8CM X 1.5CM).

 PAGE 80: SOLANA'S FLOWER, CIRCA 1997. DESIGNER UNKNOWN, SPAIN. GOSSAMER CLOTH, WIRE. 5.1" X 3.5" (12.9CM X 9CM).

 PAGE 80: PRIMAKOV'S SNOWY SCENE, CIRCA 1997. DESIGNER UNKNOWN, RUSSIA. RHODIUM-PLATED YELLOW BASE METAL, HAND-PAINTED MOTHER-OF-PEARL. 2" X 1.5" (5.2CM X 3.8CM).

 PAGE 80: VÉDRINE'S FRENCH DESIGN, 1997. DESIGNER UNKNOWN, FRANCE. RHODIUM-PLATED BASE METAL, WIRE, CRYSTALS. 2" X 2" (5CM X 5CM).

 PAGE 80: AXWORTHY'S MAPLE LEAF, 1997. ANN HAND, USA. YELLOW GOLD–PLATED BASE METAL. 1.4" X 1.7" (3.6CM X 4.3CM).

 PAGE 81: COOK'S LION, CIRCA 1970. JUDITH LEIBER, USA. YELLOW GOLD–PLATED WHITE BASE METAL. 2.2" X 2.6" (5.7CM X 6.5CM).

 PAGE 83: BUTTERFLY, CIRCA 1997. DESIGNER UNKNOWN, PALESTINIAN TERRITORIES. 18KT WHITE GOLD, DIAMONDS. 3.6" X 2.8" (9.2CM X 7CM).

 PAGE 84: PEACE DOVE, CIRCA 1997. CÉCILE ET JEANNE, FRANCE. YELLOW GOLD–PLATED BASE METAL. 2.8" X 1.7" (7.2CM X 4.3CM).

 PAGE 85: NECKLACE OF DOVES, CIRCA 1997. CÉCILE ET JEANNE, FRANCE. YELLOW GOLD–PLATED BASE METAL, LAPIS LAZULI. 25.2" (64CM).

 PAGE 86: CRAB, CIRCA 1999. VERTIGE, FRANCE. YELLOW GOLD–PLATED BASE METAL, COPPER, CRYSTALS, RESIN. 2.1" X 1.8" (5.4CM X 4.6CM).

 PAGE 87: BLACK AND WHITE TURTLE, 1990. LEA STEIN, FRANCE. CELLULOSE ACETATE LAMINATE, WHITE BASE METAL. 3.1" X 1.9" (8CM X 4.8CM).

 PAGE 87: SMALL PURPLE, BLACK, AND GOLD TURTLE, CIRCA 1980. ISABEL CANOVAS, FRANCE. YELLOW GOLD–PLATED BASE METAL, RESIN. 2.2" X 1.5" (5.7CM X 1.5CM).

 PAGE 87: LARGE PURPLE, BLACK, AND GOLD TURTLE, CIRCA 1980. ISABEL CANOVAS, FRANCE. YELLOW GOLD–PLATED BASE METAL, RESIN. 3.8" X 2.5" (9.6CM X 1.8CM).

 PAGE 87: BLACK AND BROWN RHINESTONE TURTLES, 1997. DESIGNER UNKNOWN, USA. YELLOW GOLD–PLATED BASE METAL, ENAMEL, RHINESTONES. 3.8" X 3.1" (9.7CM X 7.9CM).

 PAGE 87: BLUE RHINESTONE TURTLE, CIRCA 1998. DESIGNER UNKNOWN, USA. YELLOW GOLD–PLATED BASE METAL, ENAMEL, RHINESTONES. 5" X 1.1" (3.8CM X 2.8CM).

 PAGE 87: RED TURTLE, CIRCA 1997. DESIGNER UNKNOWN, USA. HAMMERED RHODIUM–PLATED BASE METAL, RESIN. 2.4" X 1.9" (6CM X 4.8CM).

 PAGE 87: GOLD AND LAPIS TURTLE, CIRCA 1970. DESIGNER UNKNOWN, USA. 18KT YELLOW GOLD, LAPIS LAZULI, RUBIES. 1.5" X 1.1" (3.8CM X 2.8CM).

 PAGE 88: RED BALLOON, 1992. SWAROVSKI, AUSTRIA. YELLOW GOLD–PLATED BASE METAL, ENAMEL, SWAROVSKI CRYSTALS. 1.7" X 2.8" (4.2CM X 7.1CM).

 PAGE 88: GREEN BALLOON, 1992. SWAROVSKI, AUSTRIA. YELLOW GOLD–PLATED BASE METAL, ENAMEL, SWAROVSKI CRYSTALS. 1.7" X 2.6" (4.2CM X 6.5CM).

 PAGE 88: WISE OWL, 1995. LEA STEIN, FRANCE. SIMULATED TORTOISE, CELLULOSE ACETATE LAMINATE. 1.9" X 2.4" (4.8CM X 6CM).

 PAGE 89: FISH, CIRCA 1940. NETTIE ROSENSTEIN, USA. YELLOW GOLD–PLATED SILVER, GLASS CABOCHONS, CHAIN LINK. 5.9" X 1.7" (15CM X 4.4CM).

 PAGE 89: SMALL LADYBUG, CIRCA 1995. REPRO-DUCTION, THE METROPOLITAN MUSEUM OF ART, USA. YELLOW GOLD–PLATED STERLING SILVER, ENAMEL, RHINESTONES. 1.1" X 1.3" (2.8CM X 3.2CM).

 PAGE 89: LARGE LADYBUG, CIRCA 1995. CAROLEE, USA. YELLOW GOLD–PLATED BASE METAL, RHINESTONES, ENAMEL. 1.9" X 1.6" (4.7CM X 4CM).

 PAGE 89: SWANS, 1998. SWAROVSKI, AUSTRIA. RHODIUM-PLATED BASE METAL, SWAROVSKI CRYSTALS. LARGE SWAN: 1.8" X 1.6" (4.5CM X 4CM); SMALL SWAN: 1.3" X 1.1" (3.2CM X 2.7CM).

 PAGE 90: ANGEL, CIRCA 1998. DESIGNER UNKNOWN, USA. YELLOW GOLD–PLATED BASE METAL. 2.4" X 3" (6CM X 7.5CM).

PAGE 91:
9/11 FLAG, 2001.
© DAVID YURMAN,
USA. 18KT WHITE
GOLD, DIAMONDS.
0.7" X 0.9" (1.9CM X 2.2CM).

 PAGE 92:
SUNBURST, 1987.
HERVÉ VAN DER
STRAETEN,
FRANCE. GILDED
BRASS. 3.1" X 3" (7.9CM X 7.7CM).

 PAGE 95: BEE,
CIRCA 1980.
DESIGNER
UNKNOWN, USA.
14KT PINK GOLD,
SILVER, ENAMEL, DIAMONDS,
GARNETS. 2.4" X 1.9" (6CM X
4.9CM).

 PAGE 97:
MONTICELLO
FLAG, CIRCA 1980.
BUTLER & WILSON,
UNITED KINGDOM.
RHODIUM-PLATED YELLOW BASE
METAL, CRYSTALS. 5.9" X 2.9"
(15CM X 7.2CM).

 PAGE 98:
COWGIRL HAT,
CIRCA 1999.
© ULTRA CRAFT,
USA. SILVER-TONE
BASE METAL, GOLD-TONE WIRE.
4.2" X 2" (10.7CM X 5.2CM) AND
4.3" (11CM) FROM NECK STRAP TO
TOP OF HAT.

 PAGE 100:
SPIDER, CIRCA
1995. REPRO-
DUCTION, THE
METROPOLITAN
MUSEUM OF ART, USA. STERLING
SILVER. 1.3" X 0.9" (3.4CM X
2.3CM).

 PAGE 100:
SPIDER AND
HER WEB, CIRCA
1995. REPRO-
DUCTION, THE
METROPOLITAN MUSEUM OF ART,
USA. STERLING SILVER. 2.4" X 2.4"
(6.1CM X 6.1 CM).

 PAGE 101: FLOWER
WITH FOUR BEES,
CIRCA 1940.
JOSEFF OF
HOLLYWOOD, USA.
FLOWER: YELLOW BASE METAL,
PLASTIC CABOCHON. 2.8" X 2.8"
(7CM X 7CM). BEES: YELLOW BASE
METAL. 1.2" X 1.1" (3.1CM X 2.9CM).

 PAGE 102:
MEDIUM ZEBRA,
CIRCA 1960.
CINER, USA.
YELLOW GOLD–
PLATED BASE METAL, ENAMEL,
RHINESTONES. 2.7" X 2" (6.8CM X
5CM).

 PAGE 102: LARGE
ZEBRA, CIRCA
1997. KUO, USA.
YELLOW GOLD–
PLATED BASE
METAL, ENAMEL, RHINESTONES.
4.4" X 1.2" (11.3CM X 3CM).

 PAGE 102: SMALL
ZEBRA, CIRCA
1997. DESIGNER
UNKNOWN, USA.
YELLOW GOLD–
PLATED METAL, RHINESTONES,
ENAMEL. 1.6" X 1.4" (4CM X
3.6CM).

 PAGE 104:
FOXY LADY, CIRCA
1970. LEA STEIN,
FRANCE.
CELLULOSE
ACETATE LAMINATE. 3.7" X 2.2"
(9.4CM X 5.6CM).

 PAGE 107:
AMERICAN FLAG,
2000. ROBERT
SORRELL, USA.
YELLOW GOLD–
PLATED BASE METAL, GLASS BEAD,
RHINESTONES. 6.3" X 2.8" (16CM X
7CM).

 PAGE 109: HEAR
NO EVIL, SPEAK
NO EVIL, SEE NO
EVIL, 2000. IRADJ
MOINI, USA.
TAGUA NUTS, YELLOW GOLD–
PLATED BASE METAL, SIMULATED
PEARLS, ANTIQUE GERMAN GLASS
CABOCHONS, CRYSTALS. 1.2" X
2.1" (3CM X 5.3CM).

 PAGE 110: SPACE
SHUTTLE, 1998.
RC2, CORP./ERTL,
CO., USA. TOY
MODEL MADE
INTO A PIN. WHITE BASE METAL.
2.8" X 1.9" (7.2CM X 4.9CM).

 PAGE 111:
INTERCEPTOR
MISSILE, 1998. LISA
VERSHBOW, USA.
ANODIZED
ALUMINUM. 4.1" X 1.7" (10.5CM X
4.2CM).

 PAGE 112:
SHOOTING STAR,
CIRCA 2000.
DESIGNER
UNKNOWN, USA.
10KT YELLOW GOLD, SILVER,
SYNTHETIC SAPPHIRE. 3.1" X 1.2"
(7.8CM X 3CM).

 PAGE 112:
SAPPHIRE
CRESCENT MOON,
CIRCA 1880.
DESIGNER
UNKNOWN, AUSTRIA-HUNGARY.
TIARA ELEMENT MADE INTO A PIN.
18KT YELLOW GOLD, SILVER,
SAPPHIRES, DIAMONDS. 1.6" X
1.4" (4CM X 3.5CM).

 PAGE 112:
RHINESTONE
SHOOTING STAR,
CIRCA 2000.
DESIGNER
UNKNOWN, USA. WHITE BASE
METAL, RHINESTONES. 1.9" X 0.8"
(4.9CM X 2CM).

 PAGE 112: STAR
TRAIL, CIRCA 2002.
DESIGNER
UNKNOWN,
MEXICO. STERLING
SILVER AND YELLOW BASE METAL.
5.4" X 1.8" (13.7CM X 4.5CM).

 PAGE 112:
SORCERER,
CIRCA 2000.
© Z. ALANDIA,
USA. STERLING
SILVER, FIBER-OPTIC GLASS. 3" X
1" (7.7CM X 2.6CM).

 PAGE 112:
DIAMOND STAR
AND MOON, CIRCA
2005. DESIGNER
UNKNOWN, USA.
18KT WHITE GOLD, DIAMONDS.
STAR: 1.1" X 1.1" (2.7CM X 2.7CM);
MOON: 0.8" X 0.1" (2CM X 0.3CM).

 PAGE 112: BLACK
SUN, CIRCA 2000.
DESIGNER
UNKNOWN, USA.
YELLOW GOLD–
PLATED WHITE BASE METAL,
ENAMEL, RHINESTONES. 2.4" X
2.4" (6.2CM X 6.2CM).

 PAGE 112:
MILLENNIUM
SHOOTING STAR,
2000. DESIGNER
UNKNOWN, USA.
YELLOW GOLD AND RHODIUM–
PLATED BASE METAL,
RHINESTONES. 2.4" X 1.3"
(6CM X 3.4CM).

 PAGE 113: UFO,
CIRCA 1995.
© JONETTE
JEWELRY, USA.
YELLOW GOLD–
PLATED BASE METAL, ENAMEL.
2.1" X 2.5" (5.4CM X 6.4CM).

 PAGE 114: BUG,
1997. IRADJ MOINI,
USA. YELLOW
GOLD–PLATED
BASE METAL,
CHALCEDONY, AMETHYST,
CRYSTALS, GLASS. 4" X 3.8" (10CM
X 9.7CM).

 PAGE 117:
PANTHER, 2003.
KATEL RIOU
©CARTIER,
FRANCE.
18KT WHITE GOLD, DIAMONDS,
EMERALDS, DETACHABLE LEASH.
PANTHER: 2.6" X 1.1" (6.5CM X
2.7CM); WITH LEASH: 2.6" X 1.8"
(6.5CM X 4.6CM).

 PAGE 118:
TRAILING EAGLE,
CIRCA 1963. LES
BERNARD, USA.
YELLOW GOLD–
PLATED BASE METAL, RESIN
CABOCHON, GOLD AND SILVER–
TONE CHAINS. 6.3" X 2.8" (16CM X
7CM).

 PAGE 121:
FOUNTAIN PEN,
1998. © CAROLEE,
USA. RHODIUM-
PLATED BASE
METAL, RHINESTONES. 3.3" X 0.3"
(8.5CM X 0.9CM).

 PAGE 121: MADAM
SECRETARY, 2003.
DESIGNER
UNKNOWN
(GF1W), USA.
STERLING SILVER. 2" X 1.3" (5CM X
3.4CM).

PAGE 121: HARRY S. TRUMAN, 2007. HARRY S. TRUMAN PRESIDENTIAL LIBRARY, USA. YELLOW GOLD–PLATED BASE METAL. 2" X 1.1" (5.1CM X 2.7CM).

 PAGE 122: FLOWER WITH PEARL, 2008. RUSSELL TRUSSO, USA. 18KT GOLD, YELLOW-GOLD WIRE, QUARTZ CRYSTAL, SOUTH SEA CULTURED PEARL, DIAMONDS. 2.2" X 2.2" (5.7CM X 5.7CM).

 PAGE 123: HOPE, 2008. ANN HAND, USA. YELLOW GOLD–PLATED BASE METAL. 1.4" X 0.5" (3.4CM X 1.2CM).

 PAGE 123: BARACK OBAMA INAUGURATION, 2009. ANN HAND, USA. YELLOW GOLD–PLATED BASE METAL, ENAMEL, RHINESTONES. 2" X 2" (5CM X 5CM).

 PAGE 124: KANGAROOS, 1997. ST. JOHN KNITS, USA. YELLOW GOLD–PLATED BASE METAL, ENAMEL, CRYSTALS. 1.9" X 2.1" (4.8CM X 5.3CM).

 PAGE 124: HIPPO (WITH FRIEND), 1997. ST. JOHN KNITS, USA. YELLOW GOLD–PLATED BASE METAL, ENAMEL, CRYSTALS. 1.9" X 1.4" (4.9CM X 3.6CM).

 PAGE 125: MEDIUM HOUSE OF THE NILE, 1985. AZZA FAHMY, EGYPT. SILVER, BRASS, COPPER. 1" X 1" (2.5CM X 2.5CM).

 PAGE 125: SMALL HOUSE OF THE NILE, 1985. AZZA FAHMY, EGYPT. SILVER, BRASS, COPPER. 1.1" X 0.8" (2.9CM X 2.1CM).

 PAGE 125: LARGE HOUSE OF THE NILE, 1985. AZZA FAHMY, EGYPT. SILVER, BRASS, COPPER, CORAL CABOCHON. 1.5" X 1.5" (3.8CM X 3.7CM).

 PAGE 125: GOLDEN PALM TREE, CIRCA 1997. WRA, UNITED KINGDOM; ACQUIRED IN SAUDI ARABIA. 18KT YELLOW GOLD, DIAMONDS. 2.4" X 1.1" (6CM X 2.7CM).

 PAGE 127: DRAGON AND SWORD, CIRCA 2004. DESIGNER UNKNOWN, ACQUIRED IN TURKEY. 14KT YELLOW AND WHITE GOLD, SILVER, GARNET CABOCHONS, BAROQUE FRESHWATER CULTURED PEARLS, DIAMONDS, EMERALDS. 4.8" X 0.9" (12CM X 2.3CM).

 PAGE 128: SEA HORSE, 2008. SWAROVSKI, AUSTRIA. RHODIUM-PLATED BASE METAL, SWAROVSKI CRYSTALS. 3.1" X 1.2" (7.9CM X 3.1CM).

 PAGE 128: CORAL REEF, CIRCA 2001. DESIGNER UNKNOWN, USA. GOLD-TONE BASE METAL, FRESHWATER CULTURED PEARLS, RHINESTONES. 3.4" X 2.4" (8.7CM X 6CM).

 PAGE 128: COLORFUL FISH, 2004. SWAROVSKI, AUSTRIA. RHODIUM-PLATED BASE METAL, SWAROVSKI CRYSTALS. 2" X 1.5" (5.2CM X 3.7CM).

 PAGE 128: CRAYFISH, CIRCA 2003. DESIGNER UNKNOWN, ACQUIRED IN INDIA. 18KT YELLOW GOLD, RUBIES, DIAMONDS. 2.6" X 1.4" (6.5CM X 3.6CM).

 PAGE 128: STARFISH, 2006. JOSÉ & MARÍA BARRERA, USA. YELLOW GOLD–PLATED BASE METAL, SIMULATED TURQUOISE BEADS, 3.5" X 3.5" (9CM X 9CM).

 PAGE 128: SAND DOLLAR, 2006. DESIGNER UNKNOWN, ACQUIRED IN JORDAN. HAMMERED SILVER. 2.4" X 2.1" (6CM X 5.3CM).

 PAGE 128: SEA SPONGE, CIRCA 1995. R. DeROSA, USA. YELLOW GOLD–PLATED STERLING SILVER, RHINESTONES. 1.6" X 1.6" (4CM X 4CM).

 PAGE 129: RAINBOW FISH, 2004. SWAROVSKI, AUSTRIA. RHODIUM-PLATED BASE METAL, SWAROVSKI CRYSTALS. 1.8" X 1.5" (4.4CM X 3.8CM).

 PAGE 129: SEA CREATURE, CIRCA 1997. CÉCILE ET JEANNE, FRANCE. RHODIUM-PLATED BASE METAL, CRYSTALS, ENAMEL. 2.8" X 2.8" (7CM X 7CM).

 PAGE 129: LOBSTER, CIRCA 2002. LANDAU, USA. YELLOW GOLD–PLATED BASE METAL, RHINESTONES. 5.2" X 5" (13.2CM X 12.6CM).

 PAGE 129: SEA ANEMONE, 1998. ANN HAND, USA. YELLOW GOLD–PLATED BASE METAL, RHINESTONES. 3" X 3" (7.6CM X 7.6CM).

 PAGE 129: OCTOPUS ON CORAL, 2004. KENNETH JAY LANE, USA. YELLOW GOLD–PLATED BASE METAL, RHINESTONES, SIMULATED CORAL. 2.3" X 1.6" (5.8CM X 4CM).

 PAGE 129: CHAMBERED NAUTILUS, 2006. DESIGNER UNKNOWN, ACQUIRED IN JORDAN. HAMMERED SILVER. 2.5" X 2.1" (6.4CM X 5.3CM).

 PAGE 131: LEOPARD, CIRCA 1996. DESIGNER UNKNOWN, USA. YELLOW GOLD AND RHODIUM–PLATED BASE METAL, RHINESTONES, HINGES, ARTICULATED BODY. 7.5" X 1.3" (19CM X 3.2CM).

 PAGE 131: PANDA BEAR, CIRCA 1997. DESIGNER UNKNOWN, ACQUIRED IN HONG KONG. YELLOW GOLD AND RHODIUM–PLATED BASE METAL, ENAMEL, RHINESTONES. 2.2" X 2.6" (8CM X 5.4CM).

 PAGE 131: HORSE, CIRCA 1997. DESIGNER UNKNOWN, USA. YELLOW GOLD–PLATED BASE METAL, ENAMEL, RHINESTONES. 2.2" X 2.6" (5.5CM X 6.5CM).

 PAGE 131: LEOPARD HEAD, CIRCA 1960. CINER, USA. YELLOW GOLD–PLATED BASE METAL, ENAMEL, ONYX CABOCHONS, RHINESTONES. 2.2" X 1.6" (5.6CM X 4CM).

 PAGE 131: BLACK PANTHER, CIRCA 2002. DESIGNER UNKNOWN, USA. ENAMELED BASE METAL, RHINESTONES, ARTICULATED TAIL. 3.1" X 1.6" (8CM X 4CM).

 PAGE 131: DALMATIAN, CIRCA 1996. DESIGNER UNKNOWN, USA. YELLOW GOLD AND RHODIUM–PLATED BASE METAL, RHINESTONES, ENAMEL. 2" X 1.3" (5CM X 3.4CM).

PAGE 131: YEAR OF THE PIG, 2007. DESIGNER UNKNOWN, USA. YELLOW GOLD–PLATED BASE METAL, ENAMEL, RHINESTONES. 1.9" X 1" (4.7CM X 2.5CM).

 PAGE 131: TEDDY BEAR, CIRCA 1999. CAROLEE, USA. YELLOW GOLD AND RHODIUM–PLATED BASE METAL, ENAMEL, RHINESTONES, ARTICULATED ARMS AND LEGS. 1.6" X 0.8" (4.1CM X 2.1CM).

 PAGE 132: CELEBRATION OF FREEDOM, 1998. DESIGNER UNKNOWN, USA. YELLOW AND WHITE BASE METALS, ENAMEL, SIMULATED SEED PEARLS. 2.9" X 2.3" (7.3CM X 5.8CM).

 PAGE 134: SEAL OF THE PRESIDENT OF THE UNITED STATES, 1997. BILL CLINTON'S SIGNATURE ENGRAVED ON BACK. THE WHITE HOUSE, USA. YELLOW GOLD–PLATED BASE METAL. 1.2" X 1.2" (3CM X 3CM).

 PAGE 135: ODE TO U.S. ARMED FORCES, 1998. MINA LYLES, USA. YELLOW GOLD AND RHODIUM–PLATED BASE METAL, CUBIC ZIRCONIA, GLASS, BUTTONS, ENAMEL. 2.5" X 2.8" (6.3CM X 7CM).

 PAGE 137: SILVER TRUMPET, CIRCA 1994. DESIGNER UNKNOWN, USA. SILVER. 1.7" X 0.6" (4.4CM X 1.5CM).

 PAGE 137: SILVER FRENCH HORN, CIRCA 1994. DESIGNER UNKNOWN, USA. SILVER. 2.4" X 1.2" (6.1CM X 3CM).

 PAGE 137: SILVER TROMBONE, CIRCA 1994. DESIGNER UNKNOWN, USA. SILVER. 1.6" X 0.6" (4.1CM X 1.4CM).

 PAGE 137: AMBER CELLO, CIRCA 1996. KEITH LIPERT GALLERY, USA. STERLING SILVER, AMBER. 2.8" X 1" (7.2CM X 2.5CM).

PAGE 137: RHINESTONE GUITAR, CIRCA 1994. DESIGNER UNKNOWN, USA. RHODIUM-PLATED BASE METAL, RHINESTONES. 2.5" X 0.9" (6.4CM X 2.4CM).

 PAGE 137: AMBER ELECTRIC GUITAR, CIRCA 1996. KEITH LIPERT GALLERY, USA. STERLING SILVER, AMBER. 2.8" X 1" (7.2CM X 2.5CM).

 PAGE 137: AMBER TRUMPET, CIRCA 1996. KEITH LIPERT GALLERY, USA. STERLING SILVER, AMBER. 2.1" X 0.7" (5.4CM X 1.7CM).

 PAGE 137: AMBER SAXOPHONE, CIRCA 1996. KEITH LIPERT GALLERY, USA. STERLING SILVER, AMBER. 2" X 0.7" (5CM X 1.8CM).

 PAGE 137: SILVER SAXOPHONE, CIRCA 1994. DESIGNER UNKNOWN, USA. SILVER. 1.7" X 0.6" (4.4CM X 1.5CM).

 PAGE 137: AMBER PIANO, CIRCA 1996. KEITH LIPERT GALLERY, USA. STERLING SILVER, AMBER. 1.2" X 0.8" (3.1CM X 2.1CM).

 PAGE 138: SANTA FE EAGLE, 2007. CAROL SARKISIAN, USA. YELLOW GOLD–PLATED BRASS, 23KT GOLD AND SILVER LEAF, CULTURED PEARLS, RHINESTONES, GLASS BEADS. 3.3" X 1.8" (8.3CM X 4.6CM).

 PAGE 138: TURQUOISE SANTA FE EAGLE, 2008. CAROL SARKISIAN, USA. YELLOW GOLD–PLATED BRASS, 23KT GOLD AND SILVER LEAF, TURQUOISE, CULTURED PEARLS, CORAL, RHINESTONES, GLASS BEADS. 3.3" X 1.8" (8.3CM X 4.6CM).

 PAGE 139: EAGLE, CIRCA 1940. JOSEFF OF HOLLYWOOD, USA. YELLOW BASE METAL, RHINESTONES. 3.2" X 1.8" (8.2CM X 4.5CM).

 PAGE 140: LEAF, 1998. HELEN SHIRK, USA. STERLING SILVER, 14KT YELLOW GOLD. 4" X 1.4" (10.2CM X 3.5CM).

 PAGE 141: LIBERTY, 1997. GIJS BAKKER, NETHERLANDS. STERLING SILVER, STAINLESS-STEEL WATCHES. 4.2" X 3.6" (10.7CM X 9.2CM).

 PAGE 142: SNAKE, 2005. KENNETH JAY LANE, USA. RHODIUM AND BLACK ENAMEL–FINISHED BASE METAL, RHINESTONES. 4.4" X 3.3" (11.2CM X 8.3CM).

 PAGE 143: LION, 1968. KENNETH JAY LANE, USA. YELLOW GOLD–PLATED BASE METAL, GLASS CABOCHONS, RHINESTONES. 2.1" X 3.1" (5.4CM X 8CM).

 PAGE 144: CHICK, 1994. © TIFFANY & CO., USA/FRANCE. 20KT YELLOW GOLD, RUBIES, DIAMOND, EMERALDS. 1.3" X 0.7" (3.3CM X 1.8CM).

 PAGE 144: CHICK ON BRANCH, 1994. © TIFFANY & CO., USA/ITALY. 18KT YELLOW AND WHITE GOLD, DIAMONDS, SAPPHIRES. 1" X 0.8" (2.5CM X 2CM).

 PAGE 144: SHAMAN BEAR, 2008. CAROLYN MORRIS BACH, USA. 18KT YELLOW GOLD WITH 22KT GOLD PLATING, SILVER, FOSSILIZED IVORY, COPPER. 3.1" X 2.6" (8CM X 6.5CM).

 PAGE 145: JEWELED DRAGONFLY, CIRCA 1996. DESIGNER UNKNOWN, ACQUIRED IN CROATIA. YELLOW GOLD–PLATED BASE METAL, CRYSTALS. 2.6" X 3.4" (6.5CM X 8.7CM).

 PAGE 145: BLUE DRAGONFLY, CIRCA 1997. DESIGNER UNKNOWN, ACQUIRED IN FRANCE. BLACK LACQUERED BASE METAL, ENAMEL, CRYSTALS. 2" X 1.5" (5CM X 3.8CM).

 PAGE 145: EN TREMBLANT DRAGONFLY WITH PEARL, CIRCA 1997. HEIDI DAUS, USA. OXIDIZED YELLOW BASE METAL, SIMULATED PEARL, CRYSTALS. 2.5" X 3.7" (6.3CM X 9.5CM).

 PAGE 145: YELLOW DRAGONFLY, 2000. SWAROVSKI, AUSTRIA. YELLOW GOLD–PLATED STERLING SILVER, SWAROVSKI CRYSTALS. 3.4" X 2.5" (8.7CM X 6.3CM).

 PAGE 145: SILVER MOSQUITO, CIRCA 1993. SUSAN SANDERS, USA. SILVER, ARTICU-LATED WINGS AND LEGS. 4.7" X 5.5" (12CM X 14CM).

PAGE 145: SILVER DRAGONFLY, CIRCA 1998. DESIGNER UNKNOWN, ACQUIRED IN ISRAEL. SILVER. 2.8" X 2" (7CM X 5CM).

 PAGE 145: TURQUOISE ENAMEL DRAGONFLY, CIRCA 1960. CINER, USA. YELLOW GOLD–PLATED STERLING SILVER, ARTICULATED WINGS, GARNETS, CUBIC ZIRCONIA, ENAMEL. 4.4" X 2.6" (11.2CM X 6.5CM).

 PAGE 146: BLACK SPIDER, CIRCA 2002. DESIGNER UNKNOWN, USA. YELLOW GOLD–PLATED BASE METAL, GLASS. 3.2" X 3.9" (8.2CM X 10CM).

 PAGE 146: RED SPIDER, CIRCA 2000. DESIGNER UNKNOWN, ACQUIRED IN BOTSWANA. YELLOW GOLD–PLATED BASE METAL, ENAMEL. 2.6" X 2.2" (6.5CM X 5.6CM).

 PAGE 146: PURPLE SPIDER, CIRCA 1997. DESIGNER UNKNOWN, USA. YELLOW GOLD–PLATED BASE METAL, GLASS. 2" X 2.1" (5CM X 5.3CM).

 PAGE 146: SPIDER WALIA STICKPIN, 1995. © JEWELRY 10, USA. YELLOW GOLD–PLATED BASE METAL, CERAMIC, GLASS. 3" X 3.7" (7.7CM X 9.3CM).

 PAGE 146: GREEN AND SILVER SPIDER, CIRCA 1997. E. SPENCE, USA. STERLING SILVER, MALACHITE CABOCHON. 3.7" X 3.2" (9.4CM X 8.2CM).

 PAGE 146: PEARL SPIDER, CIRCA 1997. DESIGNER UNKNOWN, USA. STERLING SILVER, RHINESTONES, MARCASITE, SIMULATED PEARLS. 4.1" X 1.5" (10.5CM X 4CM).

 PAGE 146: GREEN GLASS SPIDER, CIRCA 2000. DESIGNER UNKNOWN, ACQUIRED IN BOTSWANA. YELLOW GOLD–PLATED BASE METAL, GLASS. 1.6" X 2.4" (2.4CM X 1.4CM)

 PAGE 146: SPIDERWEB WITH SPIDER, CIRCA 1994. DESIGNER UNKNOWN, USA. YELLOW GOLD–PLATED BASE METAL, RHINESTONES. 2.3" X 2.4" (5.8CM X 6CM).

 PAGE 147: GREEN CRYSTAL FROG, 1999. DESIGNER UNKNOWN, ACQUIRED IN FRANCE. OXIDIZED BASE METAL, CRYSTALS. 2.5" X 2.2" (6.4CM X 5.6CM).

 PAGE 147: CLOISONNÉ FROG, CIRCA 1997. DESIGNER UNKNOWN, ACQUIRED IN CHINA. YELLOW GOLD–TONE BASE METAL, ENAMELED CLOISONNÉ, RHINESTONES. 2.6" X 2.1" (6.7CM X 5.4CM).

 PAGE 147: CROUCHING GREEN AND GOLD FROG, 1970. KENNETH JAY LANE, USA. YELLOW GOLD–PLATED BASE METAL. 1.9" X 1.6" (4.8CM X 4.1CM).

 PAGE 149: ROSE DE NOËL, 1970. VAN CLEEF & ARPELS, FRANCE. 18KT YELLOW GOLD, DIAMONDS, WHITE MOTHER-OF-PEARL. 2.2" X 2.2" (5.5CM X 5.5CM).

 PAGE 149: MOONSTONE DANDELION PUFF, CIRCA 1930. MAUBOUSSIN, FRANCE. 18KT YELLOW AND WHITE GOLD, PLATINUM, MOONSTONES, DIAMONDS, DEMANTOIDS. 3.1" X 2" (8CM X 5CM).

 PAGE 150: VEGETABLE MAN, CIRCA 2006. DESIGNER UNKNOWN, ACQUIRED IN BELGIUM. RESIN, CRYSTAL. 4.3" X 2.8" (11CM X 7.2CM).

 PAGE 150: SPRING ONION, 2002. MICHAEL MICHAUD/SILVER SEASONS, USA. PATINATED BRONZE, FRESHWATER CULTURED PEARL. 4.3" X 2.4" (10.8CM X 6.1CM).

 PAGE 151: APPLE, CIRCA 2006. DESIGNER UNKNOWN, ACQUIRED IN BELGIUM. RESIN. 3" X 1.9" (7.5CM X 4.8CM).

 PAGE 151: GOLD LEAF WITH RED BERRIES, CIRCA 1997. CÉCILE ET JEANNE, FRANCE. YELLOW GOLD–PLATED BASE METAL, GLASS BEADS. 2.2" X 2" (5.7CM X 5CM).

 PAGE 151: MOUNT VERNON CHERRIES, 1995. MICHAEL MICHAUD/GEORGE WASHINGTON'S MOUNT VERNON LADIES ASSOCIATION, USA. PATINATED BRONZE, DYED CRYPTOCRYSTALLINE QUARTZ, ARTICULATED CHERRIES. 2" X 2.6" (5CM X 6.7CM).

 PAGE 151: RED GRAPES, CIRCA 1995. DESIGNER UNKNOWN, USA. GOLD-TONE BASE METAL, GLASS. 2.4" X 2.8" (6CM X 7CM).

 PAGE 151: POMEGRANATE, CIRCA 2006. CILÇA, ACQUIRED IN BELGIUM. RESIN ON WIRE FRAME. 4.1" X 1.2" (10.5CM X 3CM).

 PAGE 151: THREE CHERRIES, CIRCA 1996. I. CHASE, USA. RESIN, CORD. 3.1" X 1.6" (8CM X 4.1CM).

 PAGE 151: BLACK CHERRIES, CIRCA 2006. CILÇA, ACQUIRED IN BELGIUM. RESIN. 3.1" X 2.4" (8CM X 6CM).

 PAGE 151: CLUSTER OF GRAPES, CIRCA 1990. BETTINA VON WALHOF, USA. OXIDIZED BASE METAL, RHINESTONES, ARTICULATED LEAVES. 5.3" X 4.3" (13.5CM X 11CM).

 PAGE 152: MUSHROOMS, 2000. © MARY EHLERS, USA. 18KT YELLOW-GOLD PLATING ON COIN SILVER, DIAMOND. 2.4" X 1.3" (6.1CM X 3.2CM).

 PAGE 152: HIGH-HEELED SHOE, CIRCA 2002. DESIGNER UNKNOWN, USA. YELLOW GOLD–PLATED BASE METAL, ENAMEL, RHINESTONES. 4.3" X 2" (11CM X 5CM).

 PAGE 153: SAILING SHIPS, CIRCA 2004. DESIGNER UNKNOWN, FRANCE. 18KT YELLOW GOLD, PLATINUM, DIAMONDS, ENAMEL. BLUE: 1.3" X 1.7" (3.3CM X 4.3CM); ORANGE: 1.1" X 1.7" (2.8CM X 4.2CM); GREEN: 1.7" X 1" (4.2CM X 2.6CM).

 PAGE 155: BEJEWELED MICKEY, 1989. MICKEY MOUSE © DISNEY ENTERPRISES, INC., USA. LEFT FOOT MARKED DISNEY; RIGHT FOOT MARKED CEBULLY, GERMANY. PAINTED RUBBER WITH SIMULATED PEARL, RHINESTONES, BELLS. 3.9" X 3" (10CM X 7.5CM).

 PAGE 157: GRASSHOPPER, 2001. LANDAU, USA. YELLOW GOLD–PLATED BASE METAL, GLASS CABOCHONS, RHINESTONES, ENAMEL. 3.3" X 1.2" (8.3CM X 3CM).

PAGE 157: CICADA, 1995. IRADJ MOINI, USA. YELLOW GOLD–PLATED BASE METAL, OBSIDIAN, ONYX, DIOPSIDE CABOCHONS, TURQUOISE, MOTHER-OF-PEARL, GLASS CABOCHONS. 2.6" X 2" (6.7CM X 5CM).

 PAGE 157 FLY WITH PEARL, 1997. IRADJ MOINI, USA. RHODIUM-PLATED BASE METAL, BAROQUE SIMULATED PEARL, GLASS CABOCHONS, CRYSTALS. 2.1" X 1.7" (5.3CM X 4.3CM).

 PAGE 157: GREEN LADYBUG, CIRCA 1970. © SANDOR CO., USA. YELLOW GOLD–PLATED BASE METAL, ENAMEL. 1.2" X 0.9" (3CM X 2.3CM).

 PAGE 157: TWO BLUE HORSEFLIES, CIRCA 1997. DESIGNER UNKNOWN, ACQUIRED IN FRANCE. BLACKENED AND OXIDIZED–COPPERTONE BASE METAL, ENAMEL, CRYSTALS. 1.9" X 0.7" (4.8CM X 1.8CM).

 PAGE 157: GREEN, PURPLE, AND BLUE BEETLE, 1970. KENNETH JAY LANE, USA. YELLOW GOLD–PLATED WHITE BASE METAL, GLASS. 1.9" X 1.6" (4.9CM X 4.1CM).

 PAGE 159: CO₂, 2008. STEFANIE RAHMSTORF, GERMANY. STAMPED WITH A SPECIFICALLY-NUMBERED TON OF CO_2 TO BE RETIRED. STERLING SILVER, SALTWATER CULTURED PEARLS. 1.3" X 1.3" (3.2CM X 3.2CM).

 PAGE 159: POLAR BEAR, 2000. LEA STEIN, FRANCE. CELLULOSE ACETATE LAMINATE. 3" X 1.9" (7.5CM X 4.8CM).

 PAGE 160: KATRINA PIN, 1994. DESIGNER UNKNOWN, USA. 18KT WHITE GOLD, AMETHYSTS, DIAMONDS. 2.8" X 1.8" (7CM X 4.5CM).

 PAGE 161: WRAPPING UP BOW, CIRCA 1990. DESIGNER UNKNOWN, RUSSIA. 14KT YELLOW, PINK, AND WHITE GOLD; CRYSTALS. 2" X 0.7" (5.1CM X 1.7CM).

 PAGE 162: BLACK-EYED SUSAN, CIRCA 1960. © SANDOR, USA. ENAMELED GOLD-PLATED BASE METAL. 3.2" X 1.5" (8CM X 3.7CM).

 PAGE 162: DANDELION DIAMOND PUFF, 2006. © McTEIGUE & McCLELLAND, USA. 18KT WHITE AND YELLOW GOLD, DIAMONDS, ENAMEL. 4.4" X 1.3" (11.3CM X 3.3CM).

 PAGE 162: DANDELION, 2000. © McTEIGUE & McCLELLAND, USA. 18KT YELLOW GOLD, ENAMEL. 4.4" X 1.3" (11.3CM X 3.3CM).

 PAGE 162: LILY OF THE VALLEY, CIRCA 2006. DESIGNER UNKNOWN, ACQUIRED IN BELGIUM. YELLOW GOLD–PLATED BASE METAL, GLASS, CRYSTALS, SIMULATED PEARLS. 0.4" X 1.3" (1.1CM X 3.4CM).

 PAGE 162: TULIP, CIRCA 2006. DESIGNER UNKNOWN, ACQUIRED IN THE NETHERLANDS. ENAMEL ON COPPER. 4.4" X 1.1" (11.3CM X 2.8CM).

 PAGE 163: IRISH THORN, 2007. MICHAEL MICHAUD/SILVER SEASONS, USA. PATINATED BRONZE, FRESHWATER CULTURED PEARLS. 3.2" X 1.5" (8.1CM X 3.9CM).

 PAGE 163: SUNFLOWER, 1995. © CAROLEE, USA. YELLOW GOLD–PLATED BASE METAL, RHINESTONES. 3.7" X 1.8" (9.5CM X 4.6CM).

 PAGE 163: PEARL FLOWERS, CIRCA 1995. JJ, USA. STERLING SILVER, MARCASITE, SIMULATED PEARLS. 2" X 2" (5.1CM X 5.1CM).

 PAGE 163: GOLD AND AQUA FLOWER, CIRCA 1950. DESIGNER UNKNOWN, USA. YELLOW GOLD–PLATED STERLING SILVER, RHINESTONES. 3.9" X 1.8" (10CM X 4.5CM).

 PAGE 175: ARK WITH DOVES, 2000. LANGANI/KEITH LIPERT GALLERY, GERMANY/USA. GOLD-TONE BASE METAL, MOTHER-OF-PEARL, RESIN. 3.5" X 2.2" (8.8CM X 2.2CM).

 PAGE 176: ANTS, CIRCA 1997. DESIGNER UNKNOWN, ACQUIRED IN ZIMBABWE. SILVER, CERAMIC. 0.6" X 0.4" (1.6CM X 1CM).

PHOTOGRAPHY CREDITS

ACKNOWLEDGMENTS

My world consists primarily of ideas and policies that I convey through speeches and the printed word. This book is a departure. Ideas and words are still present, but the primary means of expression is visual. The pages are graced by works of art, small sculptures in the form of jewelry. I am thankful to the designers, manufacturers, photographers, vendors, and museums who have given me, and all of us, the opportunity to enjoy these treasures.

Books, like diplomacy, require a team. They also depend on financial resources. I have long worn St. John Knits' beautiful clothing around the world, and when that company offered to help with the sponsorship of the book, I knew it was a perfect fit. I am deeply appreciative to everyone at St. John Knits for their generous support.

While some teammates are new, others are familiar. This is my fourth book since the end of my tenure as secretary of state. On each, Elaine Shocas, Bill Woodward, and Richard Cohen have played pivotal roles. Elaine, in particular, was the inspiration and driving force behind this project. Without her, there would be no book or exhibit. I have often said that she has superb judgment and perfect political pitch, now matched by her remarkable creativity. Bill Woodward, a skeptic by nature, agreed to help with the writing, though he is more comfortable with issues of war and peace than jewelry. His work helps the stories on these pages to sparkle. Richard Cohen, my editor, continues to teach me that, in writing, less is often more, while I have almost convinced him that men should wear brooches. An Olympic fencer and author, Richard is writing a book about the Sun, a task worthy of my Atlas pin.

Even a strong team requires expert help. For *Read My Pins*, I turned to Vivienne Becker, a renowned jewelry historian, author, and journalist who provided a wealth of research, important suggestions, and corrections to the text. She helped place my collection in a broader historical context. John Bigelow Taylor's photography is artistry at its best, and his images of the pins are spectacular, as are Dianne Dubler's photo compositions. Together, they gave life to my pins, brilliantly capturing a range of moods, from sorrowful to playful. Credit for the imaginative and elegant book design belongs to Rita Jules and Miko McGinty; I will not forget their patience with my numerous suggestions and countersuggestions. Diana Walker's cover photo proves that a true artist can do marvelous things, even with limited materials.

The production of this book, overseen by Melcher Media, benefited greatly from Charles Melcher's impeccably high standards and the rigorous editorial guidance and ingenuity of Lindsey Stanberry and David Brown. I am fortunate that Lindsey and David, both extraordinarily talented and seemingly indefatigable, touched every aspect of this book. Assembling a volume with so many images is complicated, but Kurt Andrews performed his magic, and Bonnie Eldon marshaled all the moving parts. It has been a genuine pleasure working with the entire Melcher team, including Duncan Bock, Frances Coy, Daniel del Valle, Heidi Ernst Jones, Coco Joly, Lauren Nathan, Christopher Nesbit, Richard Pettruci, Lia Ronnen, Holly Rothman, Jessi Rymill, Morgan Stone, Shoshana Thaler, Anna Thorngate, Anna Wahrman, and Megan Worman. I always intended that this book should be fun, and I thank them for ensuring it was.

Since 2004, my publishing home has been HarperCollins. This project was a departure for the company as much as it was for me, and I am especially indebted to executive editor Tim Duggan for his encouragement. I am appreciative, as well, of the support provided by the HarperCollins family, including Brian Murray, Michael Morrison, Jonathan Burnham, Kathy Schneider, Tina Andreadis, Kate Blum, and Andrea Rosen. The brilliant Jane Friedman was an enthusiastic early backer of this book and a source of excellent ideas. Heartfelt thanks are also due to my matchless attorneys, Robert Barnett and Deneen Howell, who have loved this project from its inception. They have earned their pin-stripes.

The Museum of Arts and Design (MAD) is a true gem. Through its collections, presentations, and educational

Ark with Doves, Langani/Keith Lipert Gallery.

programs, it celebrates arts, crafts, and design. The Museum challenges us to look at distinctive objects in an extraordinary light, thus providing the perfect setting for my pins. David McFadden, chief curator, and Dorothy Globus, curator of exhibitions, came to my home and pored over the collection. When they declared that it was suitable for Museum display, we had the green light we needed to proceed. I am enormously grateful to them both for their confidence and, more especially, to David for his thoughtful introduction to this book, and to Dorothy for helping to select the right pins and beautifully displaying them. My gratitude also goes to MAD's director, Holly Hotchner, development director Ben Hartley, and every member of the Museum's staff. Finally, the exhibit would not have been possible without the timely and generous support of Bren Simon, a talented businesswoman, a great patron of the arts, and my good friend.

This volume's impressive index of pins was primarily the work of Martin Fuller, who spent hours in my home appraising the collection for the exhibit. I assured him that the pindex would prove a labor of love, and so it proved, but with emphasis on the "labor." Marty's expertise and sense of humor both came in handy, as did the skills of those who assisted him: Colette Fuller, Audrey Hagedorn, Joanna Smith, Katharine Taylor, Lois Berger, Brenda Forman, and Marie Dotson. I also thank Diana Phillips for the early working photographs and Reema Keswani for assembling the first catalog—my favorite pin-group description was "weapons of mass destruction."

Like some of my most cherished pins, several people are in a category by themselves. In addition to providing strategic advice, Hamilton South and Anne Reingold of the HL Group were among the earliest and most enthusiastic supporters of this project. They were the first to evaluate the collection when they came to my home and spread out the pins on my bed. I am deeply grateful for all they have done and for their friendship. Their colleagues Lynn Tesoro, Joanne Langbein, Jordan Webb, and especially Arturo Diaz also provided valuable assistance.

Patricia Syvrud of Jewelers Mutual Insurance Company has kept me up-to-date on the substantive issues of the jewelry industry, many involving foreign policy matters. My friend Bonnie Cohen had a positive answer whenever I had questions, and Helen W. Drutt English deserves credit for casting a spotlight on the connection between pins and diplomacy through the wonderful "Brooching It Diplomatically" exhibit. As with my other books, Kathy Robbins provided the best advice, and through her, the legendary Paris jeweler Joel Rosenthal, of JAR, recommended Vivienne Becker to me.

Many pins lead inevitably to many thanks. In the process of preparing this book, I learned more about the provenance of my own pins and also about the history of jewelry. I am grateful to the curators, historians, and other experts who gave generously of their time, provided images, research, or advice. Contributors from the various branches of the Smithsonian Institution include Evelyn Lieberman, director of communications and public affairs; Dr. Jeffrey Post, National Gem and Mineral Collection, and Randall Kremer, National Museum of Natural History; Lisa Kathleen Graddy, Ann Burrola, and Debra Hashim, National Museum of American History; Eileen Maxwell and Christopher Turner, National Museum of the American Indian; and Lucy Commoner, Cooper-Hewitt National Design Museum. Others deserving of credit include Yvonne Markowitz, Museum of Fine Arts, Boston; Diana Pardue, The Heard Museum; Clare Phillips, Victoria and Albert Museum; June Hargrove, University of Maryland Art

History Department; jewelry historians Diana Scarisbrick in London and Elise Misiorowski in California; researchers Emma Gieben and Andrea Wulf in London; Elizabeth Frengel, Society of the Cincinnati; Daphne Lingon, Christie's; Abby Kent Flythe, Abby Kent Flythe Fine Arts; Danusia Niklewics, Hallmark Research Institute; Ralph Destino, GIA Board of Governors/Cartier; Donna Baker, Kathryn Kimmel, and Amanda Luke, GIA; Matthew Runci, Jewelers of America; Cecilia Gardner, Jewelers Vigilance Committee; Bill Boyajian, Bill Boyajian & Associates; Renée Frank, Hélène Ribatet, Jacques Guyot, and Gaëlle Naegellen, Cartier; Stanislas de Quercize, Emmanuel Perrin, and Catherine Cariou, Van Cleef & Arpels; Annamarie Sandecki, Tiffany & Co.; Nadja Swarovski, Swarovski; David and Sybil Yurman, David Yurman; Ward Landrigan, Verdura; Christopher DiNardo, Liz Claiborne/Trifari; Christopher Sheppard, Kenneth Jay Lane; Phyllis Bergman, Mercury Ring; Patti Geolat, Jewelers Mutual Insurance Company; Santa Fe artist Carol Sarkisian; London goldsmith Kevin Coates; Jim Rosenheim, Tiny Jewel Box; Ann Hand, Ann Hand Collection; and Keith Lipert, Keith Lipert Gallery. I also thank Stephanie Streett of the William J. Clinton Foundation, John Keller of the National Archives/William J. Clinton Presidential Library, James Thessin of the U.S. Department of State, Robert Pilon of the Thelonious Monk Institute of Jazz, and photographer Timothy Greenfield-Sanders.

Other colleagues and friends who helped with this project include Brandon Berkeley, Tiffany Blanchard, Laura Brent, Micaela Carmio, Kristin Cullison, Laurie Dundon, Jean Dunn, Anne Fauvre, Wini Freund, Jessy Gelber, Steven Grey, Lauren Griffin, Rachelle Horowitz, Robyn Lee, Margo Morris, Natalie Orpett, Elizabeth Raulston, Lucia Rente, Michael Ross, Karen Scates, Anna Cronin-Scott, Wendy Sherman, Jamie Smith, Jay Steptoe, Dan Sullivan, Toni Verstandig, and Fariba Yassaee. Gary Hahn deserves special thanks for enabling me to use technology as an ally in organizing and describing my collection. With her inexhaustible supply of energy, Suzy George helped to manage various parts of this project. Jen Friedman performs diplomatic magic while handling all my press and book tours. Traveling as I do inevitably leads to damage; I extend my appreciation to the Urso family at Bert's Jewelers in Washington, D.C., for keeping my pieces in good shape, especially those of the costume variety.

A book about pins requires, above all, pins. To those who have generously given me pins in the past, whether or not those gifts are displayed here, I thank you again very much. If you had any doubts about whether you chose the right gift, now you know.

Finally, as I make clear in the text, jewelry's most important role is not in diplomacy, but in the connection it establishes to loved ones. My sister, Kathy Silva, contributed greatly to this book and helped me to organize my collection and to stay (more or less) sane when I thought that one of my treasures had gone missing. I am indebted to her. Thanks to my daughters, Anne, Alice, and Katie; to my brother, John Korbel; and to every member of my family for their unwavering support and love. My six grandchildren—David, Jack, Daniel, Maddie, Benjamin, Ellie—are the real jewels in my life, and it is to them that this book is dedicated.

Ants, designer unknown.